OLIVE CULTIVATION IN ANCIENT GREECE

Olive Cultivation in Ancient Greece: Seeking the Ancient Economy

LIN FOXHALL

OXFORD
UNIVERSITY PRESS

OXFORD
UNIVERSITY PRESS

Great Clarendon Street, Oxford OX2 6DP

Oxford University Press is a department of the University of Oxford.
It furthers the University's objective of excellence in research, scholarship,
and education by publishing worldwide in

Oxford New York

Auckland Cape Town Dar es Salaam Hong Kong Karachi
Kuala Lumpur Madrid Melbourne Mexico City Nairobi
New Delhi Shanghai Taipei Toronto

With offices in

Argentina Austria Brazil Chile Czech Republic France Greece
Guatemala Hungary Italy Japan Poland Portugal Singapore
South Korea Switzerland Thailand Turkey Ukraine Vietnam

Oxford is a registered trade mark of Oxford University Press
in the UK and in certain other countries

Published in the United States
by Oxford University Press Inc., New York

First published 2007

A catalogue record for this title is available from the British Library

Library of Congress Cataloging in Publication Data

Foxhall, Lin.
Olive cultivation in ancient Greece : seeking the ancient economy / Lin Foxhall.
p. cm.
Includes bibliographical references and index.
ISBN 978–0–19–815288–0
1. Agriculture, Ancient–Greece. 2. Olive–Greece–History. 3. Olive oil–Greece–History. I. Title
S429. F69 2007
634′ .630938–dc22
2007024930

Typeset by SPI Publisher Services, Pondicherry, India

ISBN 978–0–19–815288–0

Ostinato vo' seguire
La magnanima mia impresa...
(Marco Cara, 1470–1525)

Contents

Acknowledgements

This project has grown and matured nearly as slowly as the olive trees which are its subject. Many years ago Michael Jameson and A. John Graham oversaw the inception of this research. That its development progressed when I was transplanted to Britain was thanks to the unflagging support and encouragement of John K. Davies, Peter Garnsey, Robin Osborne, and Nicholas Purcell. Much of the data and many insights and ideas that ultimately blossomed in this book were generated by my work on the Methana Project with Christopher Mee and Hamish Forbes, and on the Bova Marina Archaeological Project, which I co-direct with John Robb and David Yoon.

It is essential to thank here the institutions which have contributed to funding this research including the British Academy, the Humboldt Foundation, the Leverhulme Trust, the University of Oxford, the University of Liverpool, the University of Leicester, University College London, and Universität Freiburg. For essential information on unpublished projects I am extremely grateful to Susan Alcock, Virginia Anderson-Stojanovič, John Bintliff, Joseph Coleman Carter, William Cavanagh, John Cherry, Jack Davis, Peter Doorn, Jacques Perreault, Albert Prieto, Graham Shipley, and Anthony Snodgrass.

This work also owes a great deal to less visible but equally important support. Foremost was the ceaseless encouragement and backing of my mother, Blanche Foxhall, who did not live to see the project bear fruit, but who would have been happy to know that at last it has done so. Without the kindness, friendliness, and hospitality of the people of Koiladha and Methana in Greece, and the people of Bova and Bova Marina in Italy, this work would have withered. Helen, Anne, and Janet have been unfailingly loyal and supportive, as well as helpful and willing participants in fieldwork from a very early age, and were always there to make sure that life remained in perspective. But the fruits of this research owe most to the one with whom they were always most closely shared, to Hamish, for 'a little of everything'.

List of Figures

List of Tables

Abbreviations

A useful illustrated multilingual glossary of technical terms associated with oil and wine production can be found in Amouretti and Brun 1993: 588–95.

Aesch.	Aischines
Andoc.	Andokides
Ar.	Aristophanes
Ach.	*Acharnians*
Arist.	Aristotle
Pol.	*Politics*
[Arist.]	Pseudo Aristotle
Ath. Pol.	*Constitution of the Athenians*
Athen.	Athenaios
Cato	
RR	*De re rustica*
Columella	
RR	*De re rustica*
Dem.	Demosthenes
[Dem.]	Pseudo-Demosthenes
Hes.	Hesiod
WD	*Works and Days*
Hipp.	Hippocrates
Reg.	*Regimen*
Hom.	Homer
Il.	*Iliad*
Od.	*Odyssey*
Htd.	Herodotos
IG	*Inscriptiones Graecae*
Inschr. v. Mylasa	Blümel, W. (ed.) (1987). *Inschriften von Mylasa.* Vienna: Österreichische Akademie der Wissenschaften.

Is.	Isaios
LSJ	Liddell. H. G., Scott, R., and Jones, H. S. (1968). *A Greek–English Lexicon.* Oxford: Clarendon Press.
Lys.	Lysias
Pl.	Plato
Leg.	*Laws*
Resp.	*Republic*
Plin.	Pliny the Younger
Ep.	*Letters*
Plut.	Plutarch (references to works which are names denote biographies in Plutarch's *Lives*)
Mor.	*Moralia*
SEG	*Supplementum Epigraphicum Graecum*
SIG[3]	Dittenberger, W. (1960). *Sylloge Inscriptionum Graecarum* (3rd edn). Hildesheim: Olms.
Simon.	Simonides
Soph.	Sophokles
OC	*Oidipous at Kolonos*
Philoc.	*Philoktetes*
Theophr.	Theophrastus
Char.	*Characters*
CP	*Causes of Plants*
HP	*History of Plants*
Thuc.	Thucydides
Varro	
RR	*De re rustica*
Xen.	Xenophon
Mem.	*Memorabilia*
Oec.	*Oikonomikos*

1

Introduction

1.1. THE SCOPE

Columella (*RR* 5.8.1) called the olive 'the first of all trees', and then went on to praise its flexibility. The key point is that the tree can thrive when incorporated into many diverse agricultural and economic regimes. It can, and does, compose a major sector of many more agronomic systems than I have been able to document here. This study is centred upon the place of the olive in the agricultural regimes and economies of Classical, and to a much lesser extent, Archaic, Greece, between the sixth and third centuries BCE. The cultivation of the olive offers the opportunity to explore the intricate relationships between social and cultural values, agricultural practices, the development and adoption of technology, and the workings of the economies of Classical Greece—aspects of the ancient world which are sometimes studied in isolation from each other. This work presents a case study of one aspect of ancient agriculture at the core of economic life which is emblematic for understanding what is distinctive about economic activity in the world of Classical Greece.

Inevitably, much of the available literary and other written evidence documents the farms of wealthy land owners, especially in Attica. Among these written sources, however, some valuable epigraphical and archaeological evidence comes from elsewhere in Greece, outside Athens and Attica. Epigraphical sources raise special problems (see Ch. 2.1.3). Often it has been necessary to consider material beyond the period which constitutes my major focus, or to stray into the cultivation and processing of other crops, particularly the vine. This is particularly the case for the archaeological evidence, especially the important body of material generated by the archaeological surveys of the past thirty years, where features associated with olive cultivation and oil production can be difficult to date with any precision. Written sources, too, cover a broad time range, and their applicability outside their immediate and contemporary context is problematic. I am only too well aware that even within the comparatively narrow limits of this study there are topics that have been glossed over or ignored—hence Ptolemaic Egypt is excluded because of its distinctive ecological setting and agronomic regimes. Other sections might appear to stray far from the Classical period, or even the olive tree itself. However, there is a coherent argument about the relationships between agricultural practices and the larger economies into which they are integrated, which encompasses what might seem at first glance to constitute an eclectic selection of material.

The primary aim of this work is to explore specific olive-growing regimes in their full social, economic, political, and environmental contexts. Part of the process of understanding these contexts is to set them in a comparative chronological framework, thereby highlighting what is distinctive about how farmers in later Archaic and Classical Greek poleis utilized the olive. Assumptions about the role of the olive in Greek farming and economies have sometimes taken precedence over rigorous investigation in the scholarly literature. On occasion this has led to truth by repetition. My larger purpose, considering the olive as a kind of extended case study, is to enlarge our understanding of how specific agronomic and economic activities underpinned the functioning of Greek cities, and how they were in turn shaped by Greek social and political values.

I have not dealt with Roman olive cultivation, pressing, and processing, in any detail. This is for a different reason: from the work of Drachmann (1932) to White (1984), the phrase 'ancient oil mills and presses' has generally meant *Roman* ones. They are, consequently, well studied and documented in almost all parts of the Roman world where the vine and olive were grown, though the work of Brun (1986; 2003; 2004a; b; 2005) and of Mattingly and Hitchner (Mattingly 1988a; 1988b; 1988c; 1993; 1996a; 1996b; Mattingly and Hitchner 1993; Hitchner 1993) in Tripolitania deserves special mention. I have, however, considered various problems connected with classical Greek olive processing in some detail, since

Greek presses are less well understood, and have too often been interpreted in relation to Roman ones.

This last point is symptomatic of many studies of Greek agriculture; that many interpretations are actually extrapolations from Roman evidence. There are two serious difficulties with this methodology, which I have raised repeatedly throughout this work. The first is that the economic and political setting, and concomitantly the magnitude, of Roman agriculture is completely different from anything the classical Greek world ever saw. This is surprisingly often overlooked (Amouretti 1986; Lohmann 1993: 209–15). The question of scale had a profound effect on the whole agronomic system, right down to the pruning of olive trees and the details of propagating young plants; the impact was qualitative as well as quantitative. It also had a profound impact on the kinds of processing technologies which were adopted, and how they were employed. Certainly, it is clear that higher levels of production were achieved in the Roman world, and in Chapter 6 this is explored in more detail, though whether this increased production was always 'more efficient' is less clear.

Secondly, the olive will grow in a wide range of environmental conditions, but it does not behave the same way, and thus cannot be treated the same way, in all areas (Mattingly 1996b: 214). This is, of course, true of other cultigens as well, but the olive's very adaptability makes it easy to neglect the effects of ecological differences on cultivation strategies and techniques. So, for example, olive trees seem generally to have been kept relatively short on the farms of wealthy Greeks, while Roman Italian trees were much taller. Part of the explanation for this probably lies in the fact that in much of Italy large trees were not so likely to suffer badly from water stress as in much of southern Greece. The interplay between ecological and what one might broadly call 'cultural' factors (encompassing social, political, and economic realms) in shaping agricultural regimes is thus also a recurring theme in this study.

For my understanding of the ecological and environmental parameters I have drawn heavily on my own fieldwork experience in Greece and southern Italy, and even more heavily on the ethnographic fieldwork and sound advice of Hamish Forbes, who is frequently acknowledged in these pages. This is not to say that modern cultivation systems provide all the answers for ancient farming; most certainly they do not (Horden and Purcell 2000: 177–8), and I have pointed out many differences in the course of this work. But a firm understanding of modern farming techniques and conditions keeps one firmly aware of the limits of the possible, and brings the disparities between the present and the past into the foreground.

The concentration on the wealthiest sector of society is partly because of inherent biases in the sources; ancient 'peasants' are like post holes—you can see the places where they ought to have been, but frequently the evidence for their existence is only indirect. This is arguably the case even in the archaeological record (Pettegrew 2001; Foxhall 2001). The paucity of source material, and often its ambiguous nature, also reduces our view of what must have been a variegated socio-economic spectrum to a crude dichotomy of rich and poor. Wealthy landowners were, after all, the ones who controlled the larger share of the primary means of production—land—even in classical Athens (Foxhall 1992; 2002). This is likely to have been even more emphatically the case in non-democratic cities. Although I have tried to consider the cultivation strategies of less well-off farmers, it is very difficult to know how the other half really lived.

Finally, it has proven impossible to write an account of the techniques of olive cultivation without some analysis of the other crops with which they were grown and the economic setting of the agricultural whole. This has entailed some reconsideration of 'the ancient economy' to take account of what must after all have been the practical realities of everyday life for a large sector of the population of the ancient Mediterranean. At the heart of this analysis is the reassignment of the household to its rightful place in the explanation of ancient economic behaviour: at the front. Finley's work still forms the foundation on which much contemporary research on the economic history of the ancient world is built, although many new and exciting directions have emerged in the past fifteen years (Cartledge *et al.* 2002). But, with few exceptions (Hanson 1999), most research has started from the top down, with high-level institutions such as credit, banking, trade, labour, and all the things that economic historians of more modern periods regularly tackle. I started from the bottom up, with the practical realities and constraints of farming, and the view has proven to be quite a different one.

1.2. THE OLIVE TREE

The olive seems to be native to the Mediterranean basin. Pollen and charcoal evidence show that it was present by the mid-Holocene, well before the Neolithic period (Grove and Rackham 2001: 162). The two types of the species of olive now common in the Mediterranean (*Olea europaea L.*) are the normal domesticated olive (*Olea europaea sativa*) and the oleaster, commonly known as the 'wild' olive (*Olea europaea oleaster* or *Olea europaea sylvestris*). The two are completely inter-fertile, so although the olive and the oleaster may represent different varieties, it is unjustified to distinguish them at species level (Pansiot and Rebour 1961: 21; Forbes and Foxhall 1978; Mattingly 1996b: 216; Grove and Rackham 2001: 49). Most of the so-called 'wild' olives seen in the Mediterranean today are probably only phenotypically wild. Given that 1) the olive is both wind and insect pollinated, so genetic material spreads itself over a very wide area; and 2) the tree has probably been domesticated since at least 6500 BP (approximately 5457 BCE), it is extremely likely that genotypically, there are no truly wild olives left in the Mediterranean. In classical antiquity in areas where the olive had long been cultivated it is equally likely that many, or even most, 'wild' olives were in fact genetically mixed. In this work 'wild' is always in inverted commas, to indicate this difficulty of genotypic identification.

Olive trees are very long-lived, and very resilient, especially in the face of drought. Although the tree will grow in tropical areas, it will not fruit without a cool spell in winter in which to rest. It can survive temperatures in excess of 40° C, but is severely damaged by frost at temperatures below −7° C (Pansiot and Rebour 1961: 40–1). For this reason, as well as the late start to the growing season at higher elevations, the olive does not normally thrive at high altitudes, though its precise altitudinal limits vary considerably across the Mediterranean region (Pansiot and Rebour 1961: 43–4; Mattingly 1996b: 215). On Methana, for example, olives will not do well above around 500 m while in the North African Maghreb cultivation is viable at over 1000 m (Mattingly 1996b: 215). The tree needs a great deal of light and therefore prefers southern exposures, but is not particular as regards soil (Pansiot and Rebour 1961: 44). Nitrogen, and, to a lesser extent, potassium and phosphates, are necessary for optimum fruit production, but the olive will produce crops, albeit smaller ones, even in very nutrient-poor soils. Alkaline and even brackish soils can be tolerated (Pansiot and Rebour 1961: 48–50). In the absence of sufficient water the nutrient intake from applied fertilisers is significantly reduced (Pansiot and Rebour 1961: 118). Indeed, the primary limiting factor for growth and fruit production in most areas is the availability of water.

It is possible to raise olives commercially, unirrigated, even in areas where the rainfall never exceeds 200 mm per year, though trees do better in zones of 400–600 mm rainfall per year (Pansiot and Rebour 1961: 42). In many regions where the olive is grown it is not only the dearth of rainfall, but also its unpredictability and irregularity which present problems for agriculture. Rainfall can be erratic both interannually and seasonally. Sometimes nearly the whole of the year's precipitation can come as one rainfall event, while in other years it comes as a succession of small showers which never properly penetrate the soil. Seasonal variation is just as bad. In one year most of the precipitation can occur in the autumn, while in another year there is so little that the sowing of arable crops can be delayed.[1] Rainfall is not the only factor in the water intake of olive trees. In high temperatures plants transpire more, and soil moisture evaporates more quickly, hence evapotranspiration is a more accurate index of water availability for plant growth than precipitation. Conversely dews and fogs can substantially assist water intake in marginal areas. Olives can also tolerate water which is mildly brackish, unlike many cultigens. The most crucial time for rain to aid in fruiting, however, is in September, when a decent rainfall will considerably increase the crop (Pansiot and Rebour 1961: 53). With irrigation, olives thrive.

The morphology of the tree is well adapted to the semi-arid environments in which it flourishes. Its tough, leathery, reflective leaves minimize the amount of water loss through transpiration. Its wide-spreading, shallow root system maximizes the water and nutrients absorbed over a wide area of even thin and stony soil, though cultivation encourages the development of deeper roots as well. The tree is

[1] For examples of variability in precipitation both seasonally and interannually for specific areas in the Mediterranean see Garnsey 1988: Ch. 2.

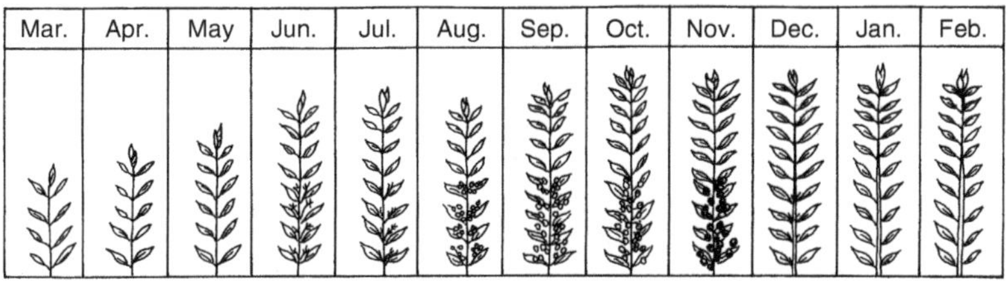

Fig. 1.1. The seasonal growth cycle of the olive tree (after Pansiot and Rebour 1961: 19).

evergreen, shedding leaves gradually throughout the year. But the main period of vegetative growth takes place between April and September, and the tree is at rest between November and March (Fig. 1.1). Flowers open in May and June and are both wind and insect pollinated. Vegetative growth is very vigorous and can occur on almost any part of the tree, even very old ones. Suckers spring readily from the roots and in tended groves must be kept at bay. Similarly, almost any part of the tree will grow roots easily, especially from cuttings and the ovules which grow at the base of the stem. Hence vegetative propagation, including grafting, is easy, and may have contributed to the early domestication of the species. Olives do not grow true to type from seed, but the seeds of domesticated trees produce phenotypically 'wild' olives (Pansiot and Rebour 1961: 16–19).

Olive trees bear fruit on two-year-old wood. The general pattern of development is that most wood growth occurs one year while fruit growth occurs the next year on the wood grown in the previous year (Fig. 1.2). At best, then, olives normally crop biennially, so often the 'on' (fruiting) and 'off' (non-fruiting) years of the tree's cycle are mentioned. Clearly there is considerable local and regional variation in the fruiting 'cycle' of the olive: in some areas, such as North Africa, the pattern may be much more complex and only roughly approximate the predominently biennial cycles normal in, for example, Greece, Italy, or Spain (Mattingly 1996b: 219–20). Though there may be some fruit produced in the 'off' years it is generally much less than in the 'on' years. If there is a really disastrous year, especially a severe drought in the 'off' year, the biennial cycle is disturbed. With no vegetative growth, in the following year, there is little or none of

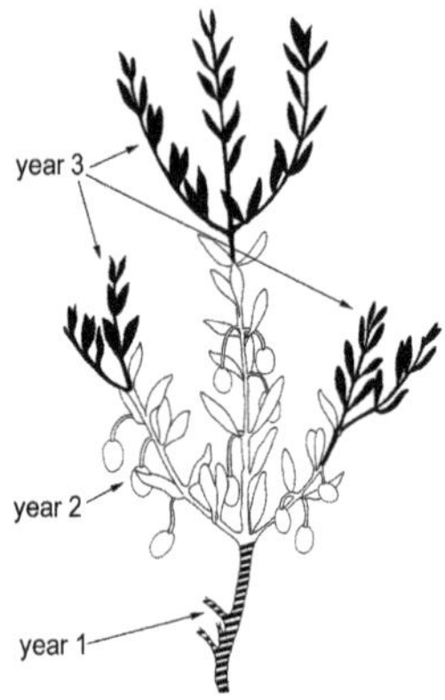

Fig. 1.2. First, second, and third year wood growth on the olive (after Pansiot and Rebour 1961: 17).

the second year wood on which most of the crop is formed. It may take several years before the trees return to their 'normal' cycle.

Generally, olive trees over a wide area are synchronized so that most bear fruit in the same year. This growth pattern is probably an adaptation to the semi-arid environments in which the tree often thrives, but it is likely there is also some biochemical or physiological cause. It is not merely the result of water stress since even heavily irrigated trees still tend to crop biennially (Hartmann and Opitz 1977: 36, 38). The biennial growth pattern is even detectable in large-scale statistical returns by country or region (Mattingly 1996b: 220; and see Fig. 1.3, where the graphs show extreme alternate year fluctuations in production both for the whole of California, where olives are irrigated, and for the whole of the Mediterranean where they are mostly dry-farmed). At present there is no way of preventing this tendency to biennial cropping, though fruit thinning, selective pruning, high levels of fertilizing and irrigation are sometimes carried out in an attempt to mitigate it (Hartmann and Opitz 1977).

Clearly, all of these factors of growth and development had a profound effect on the ways in which olives were grown in antiquity, just as they do now. Not only are there limits to what can be done with the trees themselves, but the biological parameters and growth cycle of the trees mean that they combine particularly well with some other crops and activities, but not so well with others.

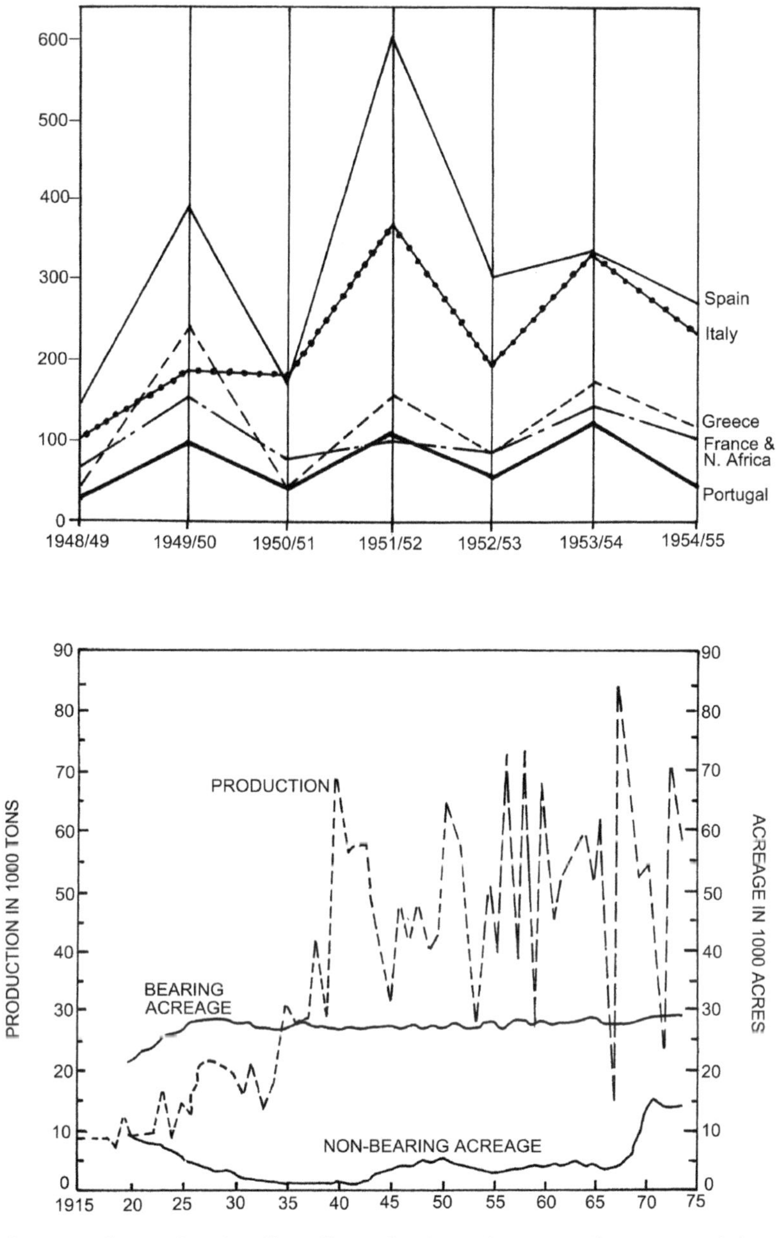

Fig. 1.3. Fluctuation in olive oil production: a) pan-Mediterranean (after Pansiot and Rebour 1961: 11); b) irrigated trees in California (after Hartmann and Opitz 1977: 6).

1.3. THE PREHISTORY OF OLIVE CULTIVATION AND PROCESSING

Where, how, and why the olive was first domesticated remains an open question, but the best early evidence appears along the Carmel coast in Israel (Galili *et al* 1993; 1997). On the underwater site of Athlit-Yam, a Late Pre-Pottery Neolithic village dating to 8100–7500 BP (approximately 7121–6387 BCE, as calibrated using CalPal Online [www.calpal-online.de]—all calibrated dates in this section were calculated using CalPal), olive pollen and carbonized olive wood are present. However, there is no evidence of olive fruit on the site, although grapes, figs, and almonds are included among the archaeobotanical remains (Galili *et al.* 1993: 152, 154). At the Kfar Samir underwater site, another early village located north of Athlit-Yam and dating to the Late Neolithic-Early Chalcolithic period, abundant remains of olive stones and evidence for processing olives for olive oil have been discovered (Galili *et al.* 1997: 1142–6). Uncalibrated radiocarbon dates on ten samples of the olive remains ranged from 6500 ± 70 to 5630 ± 55 BP (calibrated 5457 ± 68–4460 ± 63 BCE) (Galili *et al.* 1997: 1145). This suggests that the olive was domesticated some time between 7500 and 6500 BP (approximately 6387–5457 BCE). The presence of olive branches on the earlier site of Athlit-Yam suggests the possibility that domestication was discovered when a cut branch stuck in the earth for some other purpose took root. The realization that this technique could be used to select the more desirable wild forms for propagation would have effectively constituted 'domestication'. A similar phenomenon may have occurred with parthenocarpic figs at the terrestrial site of Gilgal I in the Jordan Valley (Kislev *et al.* 2006), where it has been suggested that the domestication of the fig dates as far back as 11400–11200 BP. There is no evidence in the archaeological record for the discovery of grafting, a technique which permits farmers to combine the desirable characteristics of a vigorous rootstock with a productive, more delicate fruiting scion. Historical sources make clear that it was well understood by Classical times, but how early, and where the practice began, is unknown (see

Ch. 5.7, note 2).[2] It is not easy to pinpoint a moment in the archaeological record for 'the domestication' of the olive in part because it is not possible, even from later olive remains, to distinguish cultivated from wild olives by the morphology of the stones (Hansen 1985; 1988; Runnels and Hansen 1986). In the future, DNA profiles in conjunction with large-scale detailed morphological studies may help with the identification of wild and domestic plants, as well as with understanding the spread of olive varieties, but the results of research in this area are as yet inconclusive (Terral *et al.* 2004; Elbaum *et al.* 2006).

At Kfar Samir (Galili *et al.* 1997: 1143–7), two types of features were found containing remains of olive stones. The first type (Fig. 1.4) is an unlined circular pit (D. 0.60; depth 0.50 m) dug into hard clay, with a layer of limestone pebbles at the base. The pit contained alternate layers of olive remains (olive stones, 73% of them crushed, and olive pulp) and soft organic clay with fragments of waterlogged reeds, tree branches, and straw. This appears to be the remains of a simple olive press, where crushed olives in reed frails were placed on a framework of tree branches with rocks placed on top to provide the necessary pressure. Such a 'press' would certainly have produced small quantities of olive oil, though extraction would not have been very efficient, and the process would have taken a long time. The other type of pit (Fig. 1.5) was similarly unlined and cut into the hard clay to a depth of 0.50 m deep and ranging in diameter from 0.50 to 1.00 m. These pits were filled with plant remains, mainly olive stones, most of which were crushed. They seem to represent the waste from olive crushing, which would have been useful as fuel or fodder. Large shallow stone basins, probably used for olive crushing, were also found on the site (Galili *et al.* 1997: 1145, Fig. 4).

The olive cannot be consumed by humans in its raw or untreated state. The finds at Kfar Samir demonstrate that the basic techniques of processing olives for oil are already well established here in the Late Neolithic period. However, it is intriguing to speculate on how humans came to discover the utility of the olive. One possibility is

[2] Hanson 1999: 33–5 suggests that grafting was a Greek invention, and Sallares 1991: 29–30 argues that it was only invented in the first millennium BCE but there is no sound evidence for either of these propositions.

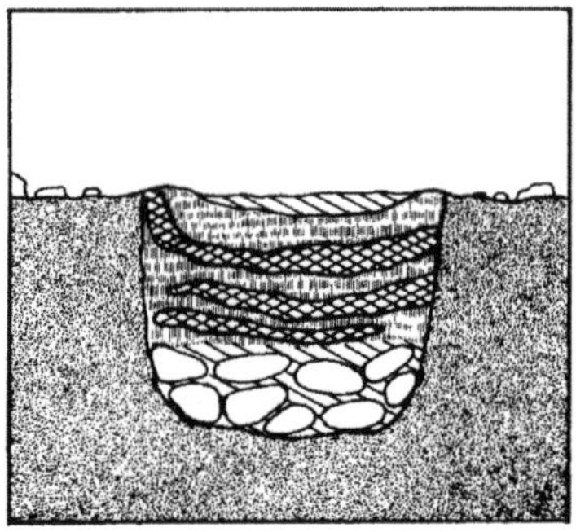

Fig. 1.4. Olive press (Installation 6) from Kfar Samir (after Galili *et al.* 1997: 1144, fig. 2a).

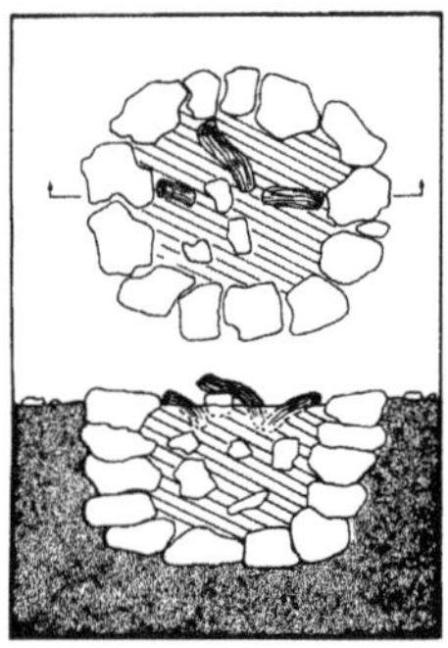

Fig. 1.5. Storage pits for waste from olive pressing from Kfar Samir (after Galili *et al.* 1997: 1144, fig. 2b, c).

that olives were first exploited for fodder, since sheep and goats would have eaten fallen wild olives when grazing. If wild olive fruit had been gathered, stored, and perhaps bruised or crushed for fodder, it is possible that oil seeping out was observed, and that ways of retrieving this oil were then deliberately sought. The simplest table olives are produced by placing olive fruits in a container between layers of salt. In these maritime communities where salt would have been readily available it is certainly possible that the table olive was also discovered (Galili *et al.* 1997: 1148), but the evidence for this is less clear. In archaeological contexts where only (or mainly) whole olive pits are found, this is most likely to represent exploitation of the fruit for table olives, not the manufacture of olive oil (Forbes and Foxhall 1978).

Closer to Greece, in the eastern Mediterranean, olive stones appear in the pre-pottery Neolithic (c. 5500 BCE) sites of Cape Andreas Kastros and Chirochitia in Cyprus (Runnels and Hansen 1986). Evidence for its domestication appears in Crete by the Late Neolithic period (fourth millennium BCE) (Grove and Rackham 2001: 162) and is claimed to occur from the early Neolithic period (fifth millennium BCE) in Spain (Terral and Arnold-Simard 1996). Less evidence for the use of the olive survives in the archaeobotanical record of mainland Greece before the late Bronze Age (Hansen 1988: 42–7; Boulotis 1996), but that may be in part an accident of preservation, since it does occur sporadically in Neolithic and Early Bronze Age assemblages. The often repeated suggestion that Late Bronze Age palaces largely exploited wild olives (Sallares 1991: 306; Hansen 1988: 46; Brun 2004a: 78), based on a speculative interpretation of Linear B ideograms (Melena 1983), is unlikely to be correct (Foxhall 1995: 241–2). By the late Bronze Age there is clear evidence for relatively large-scale olive processing on Cyprus (Hadjisavvas 1992: 3, 21–6) and in Crete (Hamilakis 1996; 1999). On the basis of our present knowledge, it is not clear whether the cultivation of the olive and the technology of its processing were diffused from a single centre or were re-invented more than once in different parts of the Mediterranean region.

1.4. TRANSFORMATIVE POWERS?

Perhaps the high cultural value, ancient and modern, placed upon the olive and the vine and their products has repeatedly led scholars to attribute special powers of social transformation to these crops. Renfrew (1972: 304–7) argued that introduction of the olive and the vine, and the 'new' technologies associated with them, were at the heart of major socio-political developments in the Early Bronze Age Aegean, and that the capacity for diversity and surplus production they provided led to complex political and social hierarchies.

Hanson (1999: 80–6) has put forward a similar kind of argument for the social and political transformations of the archaic period and the development of Greek city-states. After the collapse of the palace economies of the Late Bronze Age, in which he perceives agriculture as organised along the lines of communist collective farms (Hanson 1999: 29–30), understanding of the cultivation of vines and olives had supposedly been lost because the palaces had controlled the few domestic varieties that existed (Hanson 1999: 33–4 and notes 8–11). In any case, Hanson (1999: 32) claims, Dark Age chieftains were 'indifferent to agriculture'. He suggests that the development of 'marginal' land on hill slopes from the eighth century BCE onward, particularly utilized for 'the gradual introduction of new, improved species of vines, olives and fruit trees into Greece' (Hanson 1999: 81), came about as the result of population pressure. By the fifth century BCE, he asserts, 'we should imagine that in almost every area of Greece terraces and land reclamation were ongoing' (Hanson 1999: 80). The move onto previously under-used or unused land, he suggests, 'ensured that the agrarian ethos of Greece was transformed' (Hanson 1999: 82), both permitting and encouraging the development of the owner-occupier, citizen-farmer, whose prosperity depended on crop diversity and the surpluses produced by the exploitation of these marginal lands for wine and oil.

Also in reference to the Archaic period, Vallet's (1962) suggestion that the Greeks introduced the olive and olive oil to Italy, thus transforming Italian tastes and habits, is still surprisingly widely accepted despite evidence of the exploitation of the olive from at least the Bronze Age (Brun 2004a: 80–1; Sallares 1991: 92). Brun perceives a divide in the archaic period between the 'world of wine and oil' (located in the eastern Mediterranean) and that of 'beer, mead and animal fat'. He suggests that the cultivation of the olive and vine were lost to Greece after the Bronze Age and reintroduced from the Near East in the eighth century BCE. Even around 500 BCE, he argues, hardly any part of the Mediterranean region produced or consumed wine or oil, but the penetration of these products occurred under the influence of classical cultures (Brun 2004a: 83–8). The view that olive cultivation was (re-) discovered and spread in Greece during the first millennium BCE has been surprisingly tenacious (Sallares 1991: 305–6), despite the poor quality of the evidence to support it.

For Runnels and van Andel (1987; van Andel and Runnels 1987; Jameson *et al.* 274–5, 383–94, 400–4, 410–14) working in the Southern Argolid, the olive is a key element in the long term patterns of settlement and exploitation of the region. Their explanation is economic: at periods when the region had access to external markets, including the Late Classical, Early Hellenistic, Late Roman, and Early Modern periods, agriculture intensified and moved onto marginal lands. This was made feasible by the specialized production of olive oil for surplus and sale (with the introduction of its associated technology), and was associated with population increase and economic boom (Runnels and van Andel 303, 329). Acheson (1997) has put forward a careful and convincing critique of this rather mechanistic hypothesis, and particular aspects of it are addressed in some detail in section 6.5.3.

Were these hypotheses all true the olive (and the vine) would have wondrous powers indeed. I would argue that the olive is not a causal agent in itself. Many of the arguments proposing its transformative impact are based on rather simplistic models, some drawing heavily upon analogies with the workings of modern, market-based economics. As we have seen above (sections 1.2 and 1.3), the olive is indigenous to the Mediterranean, and appears to have been widely exploited from early times, although probably often at a very low level. The technological basics of oil production are very old, and in many circumstances small-scale oil processing is likely to be archaeologically invisible: the strongest evidence comes from underwater sites, preserving material that would long since have vanished on terrestrial sites. The olive, therefore, almost certainly played a role in virtually all agronomic regimes in the Greek world from early prehistory, and was certainly firmly established among the repertoire of cultigens well before the Archaic period. Rather, the flexibility of the olive is paramount here and the particular ways in which the olive is exploited may sometimes serve as a useful indicator to characterize and understand the larger agronomic regimes, economies and societies in which it is set. Changes in the way the olive is cultivated or technical innovations associated with its propagation or processing may signal social, political or economic transformations, but are very unlikely to have caused them.

1.5. AGRICULTURE, ARBORICULTURE, AND THE OLIVE IN EARLY GREECE

After the collapse of Bronze Age political structures, the evidence for many aspects of life in the Greek world become obscured, including agriculture. In consequence, many improbable things have been said about the subject. It is still sometimes suggested that farming diminished or disappeared and that Greece had a 'pastoral economy' in the Iron Age (Brun 2004a: 84; Hanson 1999: 32 and see section 1.4 above). Although we know little of the details of farming between 1100 and 500 BCE, in both practical and cultural terms this scenario seems highly improbable. It certainly contradicts the scanty but clear evidence that we do have (Foxhall 1995). Whatever the period to which the texts are dated, Homeric epic plainly takes farming, including tree crops, for granted. Archaeobotanical finds such as the seed and charcoal remains and pollen cores from early Iron Age Nichoria in the southwestern Peloponnese (Shay and Shay 1978) and the cereal remains from Protogeometric and Geometric Iolkos (G. Jones 1982) demonstrate that the same crops were grown in the early Iron Age as in the late Bronze Age. This does not, of course, mean that they were grown in the same way. As in preceding and succeeding periods, it appears that cereals provided the main staple food supplemented by a range of legumes, vegetables, and fruit, including grapes and olives, even if the scale of agricultural production was small by comparison. In a period of disruption, with only local levels of political organization and a relatively low population, it would be a surprise to find evidence of large-scale olive oil production, a commodity which must always have been considered valuable.

Even by the eighth century BCE it is clear from archaeological survey that the Greek landscape is hardly crowded. Throughout the Geometric and Archaic periods sites are few in number and small in size and settlement focuses on the best agricultural land (Jameson *et al.* 1994: 376; Cherry *et al.* 1991; Mee and Forbes 1997). There is no sound evidence for the exploitation of 'marginal' lands until the Classical period, and certainly no evidence that Greek landscapes suffered from over-use or over-population (Foxhall 2003; Osborne

1996b: 61–2). Indeed the main problem for farmers in Late Geometric and Archaic Greece appears to have been the problem of commandeering sufficient labour to work the land available (Foxhall 2003). Olives and vines are even more productive on good agricultural land than on rugged hillslopes. Indeed, given the range of drinking vessels and oil containers in the pottery repertoire from the Bronze Age onward (Hamilakis 1999: 47–8), there is no reason to think that vines and olives were not cultivated as part of mixed farming regimes, or were new to Greek farming.

The olive becomes particularly associated with Athens and Attica in the scholarly imagination from the seventh and early sixth centuries BCE because of its appearance in the Solonian laws. This corpus is itself problematic and in recent scholarship there is much disagreement among about how much can genuinely be attributed to 'Solon' (Blok and Lardinois 2006). The Solonian law allowing the export only of olive oil from Attica appears in Plutarch's *Life of Solon* (24), where it is stated that the law was written on Solon's first *axon*, implying that either Plutarch or his source read it (Rhodes 2006: 250–2; Gagarin 2006: 267).[3] It is difficult to understand the purpose and impact of this law since we cannot easily fit it into an historical or social context. Exempting the export of olive oil within a general ban on exports of produce from Athens underpins the idea that it was *not* a staple food, regardless of the quantities in which it was produced. In this period olive oil is likely to have been a 'semi-luxury' commodity, that is, something desirable that was sometimes within reach of people from a fairly wide band of the socio-economic spectrum, at least on special occasions, but not necessarily a staple to be taken for granted in everyday use (Foxhall 1998a; 2005; Hamilakis 1999). Table olives, in contrast, which are cheaper and easier to process, may well have been an important food. There is no archaeological evidence for the large-scale production of olive oil in

[3] The question remains open in my mind whether the wooden *axones* on which Solon's laws were written were numbered at the time of their composition and erection, or whether the numbers were added at a later date, perhaps as part of the revision and editing of Athenian laws at the end of the fifth century BCE (Lysias 30; Todd 1996). Numbering the *axones* seems out of character with what little we know of the 'publication' of early inscriptions—even as late as the first half of the fifth century, the provisions of the Great Code of Gortyn were not numbered.

seventh and sixth century BCE Attica, as one might expect to find, if it had been regularly exported in substantial quantities (compare, for example the sixth century BCE olive press at Klazomenai, Chapter 6.3.2). It is also difficult to understand how this law would have been enforced from what we know of the limited reach of the state in Athens during this period.

Attica has, however, produced the earliest Greek transport amphorae, generally known as SOS amphorae (Johnston and R. Jones 1978). It has sometimes been argued with reference to the Solonian law that these were the export containers for Athenian olive oil (Vallet 1962; Johnston and R. Jones 1978: 140–1). However, a representation of an SOS amphora on a seventh century oinochoe from a well deposit in Athens suggests an association with wine rather than olive oil (Young 1938:417 and Fig. 5) (Fig. 1.6). This is reinforced by the early sixth century François Vase, on which Dionysos is portrayed carrying one of these SOS amphorae over his shoulder (Fig. 1.7). Docter (1991; Brun 2004a: 86) has suggested that the use of these containers for wine predates the Solonian law and that thereafter they were used for olive oil. It seems more likely that these amphorae were multipurpose, and used to carry a wide range of different commodities.

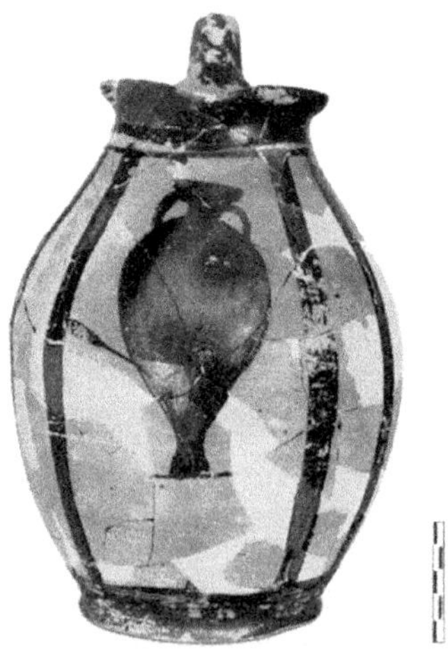

Fig. 1.6. Seventh century BCE oinochoe from a well deposit in Athens, decorated with a representation of an SOS amphora (Young 1938: 417, fig. 5, courtesy of the Trustees of the American School of Classical Studies at Athens).

Fig. 1.7. Dionysos as portrayed on the François Vase, carrying an SOS amphora, early sixth century BCE (Florence, Archaeological Museum 4209).

Our understanding of most aspects of life in Iron Age and Archaic Greece is sketchy compared with later periods. Nonetheless it seems clear that the olive, along with the other long-established Mediterranean cultigens—cereals, legumes, vines, fruit trees, vegetables—and animal husbandry, was available to farmers in Iron Age and Archaic Greece. Farmers will no doubt have cultivated and exploited it opportunistically, in ways that suited their particular circumstances and the changing environments in which they operated. These will have varied regionally as well as over time. However, there seems no need to invoke the olive (and the vine) as transforming agents in early Greek societies, or to assume that they were novelties in this period.

1.6. THE BOOK

The plan of this book moves from the general to the particular. Chapter 2 addresses the theoretical and methodological context of studying agriculture. Chapter 3 covers the problems of understanding agriculture within the context of ancient economic life and the organization of wealthy households and their agricultural holdings in classical Greece. The emphasis is inevitably on Athens, given the nature of the sources. The diversity of enterprises, agricultural and others, stands out as a major feature of this elite group. This diversity is partly an outcome, and partly a cause, of distinctive managerial and entrepreneurial structures characterized by opportunism. Chapter 4 explores the consumption of olive oil and olives in classical Greece: who, why, and how much? Although olive oil was culturally and economically highly valued, much less may well have been used than has generally been assumed. Chapter 5 analyses the cultivation of olives. The adaptability of the tree allows it to be grown within a number of different agricultural regimes. The systems of cultivation visible in Greece today are not necessarily the same as those in use in antiquity, though they may share some common features. The techniques of cultivation commonly used in classical Greece illustrate the difficulties of balancing long-term investment (as olives are) against the short-term fluctuations of a volatile, household-based economy. Chapter 6 focuses on the processing of olives for olive oil, and especially on the archaeological remains of pressing equipment. The small scale and unspecialized, modular nature of classical Greek pressing equipment revealed by both archaeological and documentary sources offers a striking contrast to that found in the Roman villas of Italy or in the deserts of North Africa. Possessing the most technologically 'advanced' equipment was not necessarily advantageous in the economic and social settings of the Greek polis. The evidence presented underpins earlier arguments about the agricultural diversity, flexibility, scale, and short-term variability which characterized the economies of classical Greece. Chapter 7 explores the role of the olive and other fruit trees in ornamental settings, highlighting another aspect of the cultural value of the olive. For the Greeks, ornamental and 'economic' plants were not distingui shed as in more modern periods. In Chapter 8 the particulars of olive cultivation are explored and how the agronomic and economic settings have become a case study for a new perspective on ancient economic life; although when it comes to distinguishing cause and effect, it is not always quite clear which is the chicken and which is the egg.

2

Wealthy Households: Theory, Sources, Methodology

2.1. THEORIES, METHODOLOGICAL ISSUES, AND SOURCES

2.1.1. Introduction

The most fundamental and difficult methodological issue facing any study of Greek agriculture is how and where to situate the political, social, and economic contexts of the source materials. Whom are we actually seeing in these sources? Whose past do they represent? There have been many answers to these intractable questions, which have resulted in widely different interpretations, most of which have contributed in their own ways to honing the terms of our investigations and improving our understanding. How individual scholars have addressed these questions has depended largely on the theoretical position (or sometimes the lack of one) from which they have started. For most, the theoretical frameworks which have formed their starting points are linked into larger, long-standing, debates about the overall character and operation of ancient economies, and the 'nature' of Greek society.

From the nineteenth century, debates about the 'primitive' or 'modern' character of economies in the classical world have been a focus of scholarship. These were brought to a head from the mid-twentieth century by the work of M.I. Finley, and, until recently, approaches generally polarized into 'primitivist' and 'substantivist' camps as opposed to 'formalist' and 'modernist' camps, although this statement is a crude oversimplification of a large and complex theoretical literature. The historiography of these debates as centred on the key figures of Weber, Polanyi, and Finley has been explored by Nafissi (2004; 2005), so there is no need to rehearse it here. Nonetheless, although most researchers working on aspects of ancient economies agree that it is time to move on, it has not always been easy to find ways to do this (Saller 2002). To some extent, we are all still working in wake of Finley and the 'primitivist/modernist', 'substantivist/formalist' debates, even as we try to develop new ways of understanding economic life and behaviour in the classical past.

In this section I will target specific issues drawn from current broader discussions of ancient economies relevant to the subject of this book and its theoretical position. Many of these issues have themselves emerged from the earlier debates. I shall then discuss the range of source material and the methodological problems of both using and contextualizing it.

2.1.2. Issues in Theory

The first issue is the problem of terminology. Miekle (2002: 248–50) points out that the language of modern economies most of us have used for discussing ancient economies is misleading. Terms such as 'capital', 'entrepreneur', 'credit', and 'risk', are not neutral. Even when ancient historians use them correctly (which sometimes they do not), such words carry meanings intrinsic to modern economies which have no counterparts in the ancient world. However, here we confront the limits of English, German, French, and other modern languages—there are no meaningful 'neutral' terms available to use. As Miekle (2002: 247–9) says, all of us, like it or not, are coming to antiquity from a world which has had modern economic institutions for a long time, and it is difficult or impossible to shed these modern meanings. There is no easy solution to this, and it is futile to invent meaningless neologisms. The best compromise seems to be not simply to acknowledge, but to highlight the problem, and to flag that 'economic' terms will carry a different range of meanings when set in the context of ancient economies—they are not synonymous with their modern counterparts but become analogous to them.

The second issue is the unacknowledged elements of the legacy left from earlier debates. It remains the case that most studies of ancient economic institutions still implicitly or explicitly contrast Us with Them (von Reden 1995: 127), in a way that, as Saller (2002) points out, leaves us open to the charge that we are simply engaging in colonialist discourses. In particular, much recent research accepts the premise that in classical antiquity the realm of the economic was to some degree 'embedded' (to use Polanyi's still much used and abused term, see Nafissi 2004: 400–5) in other kinds of culturally-determined relationships—social, political, religious, and so forth, but that this is not the case in the modern world. That we thereby run the risk of 'primitivizing' the Greeks and Romans by constructing them as Other becomes particularly evident in studies which highlight gift-exchange, reciprocity and similar kinds of social relations as vehicles for 'other' economic behaviours in antiquity (Gill *et al.* 1998; Kurke 1991; 2002; Tandy 1997: 7–8; see Foxhall 2005: 241), which supposedly do not belong to the marketplace of modernity. Of course there are many ways in which economic behaviours and institutions in the ancient world plainly differed greatly from those in the modern world, but the grounding of economies in culturally-determined frameworks is not one of them. The idea that 'the economy' or 'the market' is impersonal and 'disembedded' in our own or any other time seems to me, as a 'self-proclaimed "irredentist substantivist"' (Foxhall 2002: 219; Nafissi: 2004: 403; 2005: 270), no more than an ideal of our own society. Surely our own economy is also 'embedded', in the sense that it is structured by the culture(s) and societies in which it operates, even if as a global, mega-institution, viewed from the stratosphere of macroeconomics, it also has a kind of independent existence with its own cycles and momentum (though even these, of course, need not necessarily be independent of cultural principles or international politics). And 'market relationships', depending on the scale at which one observes them, are not always 'impersonal', nor do they necessarily function independently of social, political, and other kinds of relationships, even if we believe they should (Granovetter 2005).[1] I do not mean by

[1] At the time of writing, the British government is wrestling with the problem of a serious gender gap in pay (men earn on average 17% more than women, and many

this that ancient and modern economies are alike, merely that both are culturally grounded constructions. The difference in the impact of 'embeddedness' is to some extent the disparity in scale between ancient and modern societies and economies, in conjunction with the relative size and dominance of economic institutions in relation to other kinds of social institutions, which are plainly very much greater in the modern world. So perhaps rather than disembedding Athens (Nafissi 2004: 404–5; 2005: 261), we should 're-embed' ourselves.

Most paradigms for understanding ancient economies were generated in a period when scholars still believed in the possibility of universally applicable rules and systems, and the evolutionary development of institutions and societies over time.[2] This is not a heritage that is easy to escape. Different scholars have tried different tactics. This leads us to ask, with Miekle (2002: 245–6), though perhaps from a slightly different angle, to what extent economic 'laws' are useful for understanding ancient economies, or are we dealing with creatures of two different species?[3] Awareness of this particular aspect of our theoretical legacy is important for avoiding mechanistic, explanatory paradigms or oversimplifying by focusing on the normative, to the exclusion of significant anomalies and complex patterns of behaviour.

As for most ancient historians and archaeologists researching in this field, a large body of social theory, much of it originating in anthropology, underpins the theoretical framework of this book. The key theoretical approach of this work lies in the notion that all social, economic, and political institutions in all societies are shaped

women are over-qualified for the jobs they hold), while the present Minister of Culture, Tessa Jowell, is trying to clear herself of allegations of involvement with a very large amount of money received by her husband from contacts of the Italian Prime Minister, Sergio Berlusconi, as a 'gift'. These are certainly not isolated instances of contemporary economic relationships entangled with social and political ones, and many more examples from all over the world could be cited.

[2] For example, Miekle (2002: 236) representing the relationship of the 'modernist position' to the study of economics states; 'economics is the study of laws which hold between humans and economic goods... These are laws of such general character that they must apply in all periods of history and to all forms of society...'.

[3] Meikle 2002 couches this in terms of ancient economic institutions characterized by use—value in contrast to those of the modern world characterized by exchange value.

by—'embedded in', one might even say—their particular cultural and temporal contexts. In other words, the one generality is that there are no generally applicable predictive 'laws' for understanding and explaining social institutions. This is not to say that comparison is not possible or useful, when different societies appear to share specific features. Rather, comparison functions admirably as analogy, and is helpful for highlighting both similarities and differences between different societies. However, its predictive value, like its use, must be kept well under control. The more points of similarity and the more limited the comparison, the more useful the analogy is likely to be for understanding the full range of meanings an object, landscape or social institution might have had in the past (Hodder and Hutson 2003: 26–30, 148, 194). For a study such as this one focused on agriculture, where technical and environmental factors necessarily play a major role, analogies can be strong and particularly enlightening where we can be certain that we can match these variables on both sides of the comparison. However, when using modern 'proxy data' or comparative material, one must be aware that most societies of the nineteenth century or later are likely to be tied into larger, even global, economies, at least to some extent—a rather different setting from the relatively self-contained economies of classical Greek city states.

The debates surrounding 'embeddedness' also raise related questions about agency and engagement which touch on the theoretical background to this book. Within any particular society, not everyone will necessarily engage as actors with the realm of the economic to the same extent or in the same ways; however, that realm is constructed in any particular society and wherever it is deemed to be located. Therefore, 'economic' institutions, of whatever type or at what ever scale they exist in any particular society, will not necessarily mean the same to everyone, or function in the same way for all. In some cases this may be by choice, more often it may be by necessity or compulsion. Again, this is the case in different ways and to different degrees for both ancient and modern societies.[4]

[4] At the time of writing, comparison of the African-American families in New Orleans who faced Hurricane Katrina and its aftermath without insurance, with the circumstances of white families, strongly suggests differential engagement with economic institutions in that situation.

In this book, the object of study, the olive, offers itself as a case study well suited to exploring ancient conceptions of value. Consumption reflects the desire for, and the desirability of, valued objects. It is undisputed that throughout antiquity olive oil, especially that of perceived 'high quality', was greatly valued in both economic and in social and symbolic terms. Before the Romans, the high-volume production of olive oil was rare. Exploring patterns of production and consumption of olive oil opens the possibility of unpicking the qualities which made it desirable. It is clear that this transcends 'necessity', and thus desirability is not the same as 'demand' in its modern economic sense. Consumption was the endpoint of desire, and in Archaic and Classical Greece, the consumption of olive oil was also its embodiment, in literal terms, as food, cleansing, and adornment. Such embodiment could both reflect and construct the identity and status of the individual consumer, as part of the occasion when it was consumed (Foxhall 1998a; 2005). 'Embeddedness' can thus be linked to agency through consumption and its contexts. In contrast, the other important product of the olive tree, table olives, had a much lower value, and appears to lack the symbolic resonance of olive oil. These issues are explored in more depth in Chapter 4.1.

2.1.3. Situating the Sources

The paucity, temporal and regional fragmentation, and sometimes the sheer impenetrability of the sources have proven to be major obstacles for understanding Greek agriculture. For a start, most texts are unquestionably Athenocentric, emanating from that least typical of poleis between the fifth and third centuries BCE. Philosophical treatises, such as Xenophon's *Oikonomikos* and Theophrastos' botanical works, the surviving speeches from the law courts, the inscriptions issued by the *poletai* (the magistrates who acted as state auctioneers for the Athenians), the fourth century *horoi* (boundary stones signifying property that stood as security for a loan), and the inscribed records of leased land, constitute the greatest bulk (though by no means all) of the literary and epigraphical material relevant to Attic land use. In addition, the city has been the subject of large-scale, long-term excavations, most notably those of the American School of

Classical Studies (1953–) at the Agora and the Deutsches Archäeologisches Institut at the Kerameikos (1939–), as well as an enormous amount of rescue excavation by the Greek Archaeological Service. Parts of the countryside of ancient Attica (those at least not built over by modern Athens) have been quite thoroughly explored, and many farm sites have been recorded (J. Young 1956; Langdon and Watrous 1977; Langdon 1990–1; Wickens 1983; S. Morris and Papadopoulos 2005). More systematic surveys of the countryside, in particular the Stanford Survey in northern Attica on the Boiotian borders (Munn and Munn 1989) and Lohmann's survey in southern Attica (Lohmann 1985, 1993, 1993, 1994), have also filled out the picture. Additionally, the excavation of two country houses by the British School at Athens has provided invaluable, if not unambiguous, evidence for Attic rural life (J. Jones, *et al.* 1973, J. Jones, *et al.* 1962).

Such a bias in the evidence must obviously affect our understanding of regional variation, despite the existence of a few literary texts (notably Hesiod's *Works and Days*) and a number of relevant inscriptions from other parts of the Greek world. Inscribed leases of land and other agricultural resources present special problems. Often they emerge from very particular or unusual contexts and therefore cannot be read as 'typical' of ancient practice. They offer a good example of the difficulties since land cultivated by a tenant was probably not cultivated in the same way and with the same aims as land which belonged to a household and which was intended to be passed on to future generations. The differences between the cultivation of leased and owner-occupied land are likely to have been most different when it comes to perennial tree crops, especially long-lived ones like the olive. So, for example, this has been cited as a reason why olive cultivation seems to be rare among the Delian and Rheneian estates belonging to the sanctuary of Apollo for which detailed records survive (see ch. 6.7; Brun: 1997: 605–9; Kent 1947: 288; Reger 1994; Rauh 1993). Archaeologically documented olive presses in the city of Delos and records inscribed on stones of the purchases of olive oil by the Sanctuary officials may suggest that olives were not equally rare on private lands.

In part this regional bias has been alleviated by archaeological finds, including both excavation and regional survey, in many different parts of Greece and Italy. Excavation and survey of rural sites in the *chora* of Metaponto in Italy (Carter 1990; 2006; Prieto 2005), and in the Berbati Valley in the Peloponnese (Hjohlman *et al.* 2005) have provided valuable insights into the polis landscape. Urban sites where the settlement areas have been extensively excavated, most notably Olynthos (Robinson 1929–52; Robinson and Graham 1938), and Halieis (Jameson 1969, Rudolph 1978; Ault 1994a, 1994b; 1999), also provide important evidence. Unfortunately, other areas, no matter how well documented archaeologically, lack the sheer volume of detailed historical evidence for land use and land tenure.

Generally, the evidence for agricultural practices in Greece is fragmentary in all senses. Frequently, a picture must be built up out of information derived from several different places, many kinds of source material of varying quality, and spanning a two to three hundred year range. In consequence, it is difficult to produce a picture of nuanced change over time. And, as noted at the beginning of this chapter, it is not even clear who the farmers are that we think we see: there are radically different views about where we should situate them in their societies, and indeed even how Greek societies were stratified.

Several American-based scholars have in different ways focused on a perceived split between what they call 'middling' and 'elitist' groups and ideologies in Greek societies (Hanson 1999; I. Morris 1996; 2000; Kurke 1991; 1999; 2002). What all three of these scholars have in common is that they perceive the origins of a 'middling' group and its associated ideology in the literature of the archaic period, which then becomes associated with the democratic ideals of classical Greek cities and citizens, though they vary in the ways in which they interpret and apply these categories.

Hanson (1999) perceives the rise of the 'middling' farmer as the major trajectory for the development of Greek polis society. From the literary figure of Laertes (Hom. *Od.* 24) and the poet Hesiod's self-portrayal in the *Works and Days*, he builds a vision of a landscape populated with sturdy yeoman 'homestead farmers', who are neither poor subsistence peasants nor members of a traditional aristocratic elite (Hanson 1999: 86–8, 105–11). For Hanson (1999: 87), this was a 'Greek agrarian revolution' in which these 'middling farmers', who eventually evolved into the stable foundation of the classical polis, were egalitarian and equal—effectively a 'middle class':

> All the farmers of the neighbouring community were to hold land roughly similar in size. Among this broad agrarian class of Greek *mesoi*, land should ideally be passed down through the family without alienation. The ancestral property (*klêros*) belonged to a family (*oikos*), not a single (and unpredictable) individual to do with it as he pleased. The few holders of large fortunes, even those non-propertied rich involved in trade or mining, should be subject to a variety of restrictions and limitations, preventing accumulated capital from being expressed in large landed estates (Hanson 1999: 180).

Hanson's approach to the evidence is broad-brush and selective, and the analysis is often at a very general level. He rarely explores the difficulties of interpreting ancient texts, and he is prepared to accept at face value the information presented by classical and Roman writers about their own distant past, even when other kinds of evidence contradict this information. In particular, his use of archaeological material is limited and often second hand. At one level his vision is profoundly romantic, and his comparisons with the pioneer farmers who settled the American West with his own other life working a family farm, are overt. One is left with the feeling that this is more an exploration of modern America than of the classical world.

Kurke's (1991, 1999, 2002) important and imaginative work has investigated the relationships between coinage and money on the one hand, and the moral values expressed in archaic and classical literature, in particular archaic poetry, on the other. Drawing on the work of Parry and Bloch (1989), she associates the introduction of coinage and money both with the newly emerging civic values of the polis and with the negatively valued 'short-term transactional order', while elite values of reciprocity and gift exchange were associated with the positively valued of the 'long-term transactional order'. These elite values were thus hostile to coinage and monetary transactions. For Kurke (2002: 94), the opposing ideologies detectable in this poetry correspond to a 'middling' position (associated with civic ideology and the polis), and an 'elitist' position (associated with a traditional established aristocracy). The two positions also correspond with different genres of poetry, the 'middling' with iambic verse, and the 'elitist' with monodic lyric (Kurke 2002: 93). Both positions attempt to appropriate the 'long-term transactional order' for themselves while relegating their opponents to the 'short-term transactional order'.

The correspondence between of all these elements, however, turns out to be not quite so tidy. Von Reden (1997) has convincingly demonstrated that money and coinage were not used exclusively for exchanges which belonged to the 'short-term transactional order'. Moreover, von Reden (1997) shows that much of the archaic poetry on which the original argument (Kurke 1991) depended predated the use of coinage in Greek poleis. But, at the same time, other kinds of weighed objects were being used to establish values both in the civic organisation of the polis (e.g. fines for magistrates) and in terms of longer term reciprocal exchanges (e.g. votive dedications to the gods, prizes in athletic games). In addition, there is an implied contrast in Kurke's work between Us and Them which may not be justified; highlighting the 'Otherness' of the Archaic Greek world by implying that the long-term transactional order has no place in the modern world. There is a danger here, apparent also in subsequent secondary literature (Gill *et al.* 1998), of presenting features associated with long term transactional orders such as 'reciprocity' as explanations in their own right, rather than as the description of social behaviours which they are (Foxhall 2005: 241).

Ian Morris (1994a; 1996; 2000) has developed the notion of the 'middling' citizen-farmer most fully. Drawing on the work of political scientist Robert Dahl (1989), Morris argues that Athenian democracy was founded on a 'middling ideology' fulfilling Dahl's (1989 30–1) theory of the Strong Principle of Equality. This, Morris (2000: 111) argues, emerged in Greece in the late sixth century BCE. He constructs a model of this 'middling ideology', founded upon 'the middling culture of civic manhood' (I. Morris 2000:112). He draws upon fourth century law court speeches and philosophy to demonstrate that the 'moderate' man, who was measured, self-controlled, and respectable, was 'a powerful native model' (I. Morris 2000: 113). Speakers construe themselves and their audience (the jurors) as belonging to this group of 'middling' men, and attempt to portray their enemies as behaving in ways which put them outside this positive ideology (I. Morris 2000: 115, 120). Indeed, Athens itself is construed in these rhetorical contexts as 'a community of *metrioi*, content with a little money' (I. Morris 2000: 113). The term 'did not mean a middle class in an economic or occupational sense' to Athenians (I. Morris 2000: 115), but when applied to rich men, implied that they used their wealth for the benefit of the city. This ideology was contested and under threat from 'outsiders' (e.g. the poor, the *kinaidos*, homoerotic passive). For Morris (2000: 116), 'When Athenians called themselves *metrioi* they imagined one another as self-sufficient farmers on their own land, heads of households, married with children, pious, responsible, and self-controlled.'

As an ideology which permeates Athenian rhetoric, though perhaps not so pervasively or dominantly as Morris suggests, the 'model' of the 'middling man' is reasonably persuasive. However, having said that, it does not refer to 'middling' in an economic sense, Morris (2000: 140) proceeds to argue that land was distributed relatively equally among citizens. He cites as evidence of this analyses of Attic landholding by Osborne (1992a) and Foxhall (1992) who argue, independently of each other, that most land was in the hands of the wealthy. Morris applies the Gini coefficient to their results to show that Athenian landholding was much more egalitarian than the situation in Rome and elsewhere (I. Morris 2000: 140–4; 1994a: 362 and n. 53). Apart from the question of whether the Gini coefficient is a meaningful statistic in relation to these figures,[5] or whether a 10%

[5] The Gini coefficient is a standard measure of inequality in modern economics and other social science research. It is derived from the Lorenz curve, a statistical method in which groups of data are arranged in ranked order from largest to smallest. (A hypothetical example of such groups would be the numbers of landowners in a society with holdings of 0–1 ha, 1.1–3, ha; 3.1–5 ha, etc., or the total area of landholdings of plot holders owning 0–1 ha, 1.1–3, ha; 3.1–5 ha, etc.) Once the table is set up, each group is converted into a percentage of the total, then the cumulative percentages are calculated and plotted as a graph. A totally equal distribution would result in a diagonal line, so the more the Lorenz curve veers away from the diagonal, the less equal the distribution. The Gini coefficient is the space between the diagonal and the Lorenz curve divided by the area of the graph below the diagonal. Perfect equality equals 0, perfect inequality equals 1. For the analysis of contemporary, large-scale social science data, the Gini coefficient is a widely accepted and reliable index of inequality. Unfortunately, for the ancient Greek world we lack such large-scale quantitative data. Although there are data which might allow limited use of this statistical method in parts of the Roman world, it seems doubtful that the estimates in Foxhall (1992) and Osborne (1992a) are adequate for the use of this statistical methodology.

error is acceptably precise, Morris's arguments falter on the issue of scale. Comparing Athens with Rome is not just comparing 'apples with oranges' but a bowl of apples with a truckload of oranges. I have suggested elsewhere (Foxhall 2002: 215) that the distance between the smallest and largest landholdings were much greater in the Roman world, and I return to that issue in Chapter 8 of this work—plainly, the scale and complexity of the Roman economy was enormous compared even to Athens (let alone any other Greek polis). However, the Athenians knew about neither the Romans nor the Gini coefficient: some Athenians will simply have observed that others had larger or smaller landholdings, or greater or lesser wealth, than themselves. How they felt about it we can only guess, but the difference between a Phainippos (Dem. 42) or an Alkibiades on the one hand and a Dikaiopolis (Ar. *Ach.*) on the other might have looked very much greater from a perspective within classical Athens than it does from a modern point of view, from outside and with the benefits of a wider comparative perspective. Indeed, Morris's point that Athenian speakers portrayed themselves as 'moderate' in public contexts such as the law courts while simultaneously casting their opponents as both wealthy and selfish, using their riches for their own luxurious lifestyles, not for the benefit of the city, probably suggests that such representations were underpinned by a deep-seated belief on the part of less wealthy citizens that there really was a significant gap between their own resources and those of the stars of the oratorical show.

At one level, the difference between Ian Morris's interpretation on the one hand and those of Osborne and Foxhall on the other is no more than whether one assembles the scanty and fragmentary source material to show a glass half full of water or half empty, though other scholars have also highlighted problems with the workability of 'middling' and 'elitist' ideologies (Kistler 2004; Hall 2007: 178–81). What is much more secure, however, and casts considerable doubt on the construct of the 'middling ideology' as applied by Hanson, Kurke, and Morris, is that *all* of our literary source material was written by the wealthy from an elite perspective, even in the democratic regimes of fifth and fourth century Athens. Further, it is probably the case that inscriptions too were mostly framed by elites, even if they are, or purport to be, carrying out the wishes of the citizen body as a whole.

This is not to say that the ideology of political equality was not important in Athenian democracy. Plainly it was extremely important, and it is clear that the wealthy took considerable pains to ensure that they represented themselves to be as 'ordinary' as possible. However, my point is that we simply do not have the voices of the *hoi polloi* directly represented in our literary or, for the most part, epigraphical, sources.

The other worrying feature of this scholarly construction of the 'middling ideology' is the elision between the realms of politics and economics in aspects of life where these are demonstrably separable, in the sense that the political equality of Athenian citizens was distinct from the differentiation in status as defined by wealth. To some extent that blurring is present in the ancient sources themselves as part of the rhetoric, particularly in Athenian law court speeches where speakers sometimes wish to gloss over their own wealth and to portray themselves as modest and ordinary men in the community of the Athenians (Dem. 44.4), but more often stress the wealth and greed of their opponents (Dem. 47.78–80; Dem. 42.3–4). This is all the more important for these speakers if in Athens the ideology of political equality was perceived to be juxtaposed with considerable inequalities in wealth and power. However, these modern academic representations of a 'middling' ideology are inclined to accept this elision in the rhetoric as having some basis in reality, and we are in danger of anachronistically re-introducing a class of bourgeoisie into the political economies of ancient Greece by the back door.

If in written sources we largely hear the words of the wealthy, what about archaeological source material? Here there is no clear answer or consensus among scholars about where to situate it within a socio-economic spectrum. Archaeological survey has revolutionized our understanding of rural landscapes of the Greek work over the past thirty years. Ancient historians (Hanson 1999: 187–8; N. Jones 2004: 46) have regularly assumed that the pattern of scattered rural 'farmsteads' which has emerged from the results of archaeological survey in many parts of Greece validates the picture of a countryside populated by owner-occupiers, 'middling farmers', 'subsistence peasants', or 'sturdy yeoman hoplite citizens'. First-hand examination of the evidence allows us to jump to no such easy conclusions.

It has regularly been pointed out (Osborne 1985a; 2004; Cherry *et al.* 1991: 337–8) that the number of rural sites discovered through archaeological survey would not accommodate a very large proportion of the agrarian population of any Greek polis. Nor is it clear what the dots on the maps represent: certainly, like the written record, they should not be read at face value (Foxhall 2004b; Pettegrew 2001; Osborne 2004). It is unlikely that archaeologists have recovered all sites in the areas they have field-walked (and these areas are themselves samples from the total landscape). Guesses have been made about the proportion of total sites recovered (Cherry *et al.* 1991: 16–20), but it is generally acknowledged that such estimates are unlikely to be very accurate. The proportion of sites surviving and recovered may vary considerably in different areas depending on local geomorphological, erosional, and landscape histories (Bintliff 2005; Ayala 2005).

On the other hand, it is also fairly certain that the collection of dots on any particular survey map were not all simultaneously occupied, even within a period such as the 'Classical' or 'Hellenistic' (Foxhall 2004b). The dating of sites discovered in survey is imprecise by the nature of the exercise, and historians often seek to pin rural settlements within much more precise chronological boundaries than the archaeological evidence can bear. Nor is it certain that all 'farmstead' sites were occupied in the same way. Some may have been occupied only seasonally (Osborne 1985a; 2004: 170): the Vari House is a good candidate for this (Foxhall 2001). The Classical phase of the Berbati 'farmhouse', for example, turned out to be a pottery kiln (Hjohlman *et al.* 2005), while the tower, a feature generally associated with 'farmhouses' (S. Morris and Papadopoulos 2005), dated to the Hellenistic period. In many parts of the Greek world most farmers may have lived in towns or villages. The constant high background scatter of classical pottery on the outskirts of urban sites in Boiotia suggests intensive use and cultivation by the inhabitants of the town. Pettegrew (2001) has also suggested that such scatters might also mask the archaeologically invisible dwellings of poorer and smaller-scale farmers (but see also Bintliff *et al.* 2002).

The fact that rural sites *are* detectable as discrete sites may then suggest that they were quite substantial establishments, which were not occupied by modest small-scale farmers. Some of the tower farmhouses are clearly very substantial and it is likely that we can assign many of these to larger-scale, wealthier farmers (S. Morris and Papadopoulos 2005). Some of the few excavated examples we have also suggest fairly prosperous owners, the Dema House in particular is substantial in size and has a ceramic signature similar to an urban house (Foxhall 2004a). This may suggest it was the country home of a wealthy citizen rather than a 'farmstead'.

For Attica most scholars have assumed that 'farmsteads' were owner-occupied, and that may be correct, although doubts have been cast on this premise by S. Morris and Papadopoulos (2005) and others (Osborne 1992a). But, there is no reason to assume this for 'farmsteads' associated with other poleis in the Greek world. Can we be certain this was true for oligarchic Metapontion, for example? Superficially similar landscapes may mask radically different political economies, and the state of our knowledge does not allow us to distinguish these with certainty.

In summary, it is difficult to situate the rich evidence supplied by landscape archaeology within the historical socio-economic spectra of Greek cities. Certainly we should not assume that the small rural sites were all occupied by small-scale farmers. The same is true of agricultural equipment associated with these rural landscapes, even when we can date it securely to the Classical or Hellenistic period (see Chapter 6). Visual and iconographic evidence is equally ambiguous: many agricultural scenes on Attic vases may be ritual (for example, ploughing scenes). Practically all vintage scenes appear to be carried out by satyrs and maenads, not humans. The few left are impossible to contextualize, so, although they are interesting and important, they are also problematic. It is probable then that the archaeological record, like the textual record, tells us more about larger-scale wealthier farmers than about smaller-scale farmers, but we cannot be certain of this.

2.1.4. Wealthy Land Owners and Larger-scale Farmers in Ancient Greece

'Greek farming' has often been treated as a monolithic category in the scholarship, so the distinctive characteristics of larger-scale Greek land owners and their agricultural holdings have not generally been recognized (S. Morris and Papadopoulos 2005). Frequently, it has been assumed that farmers had similar aims at all levels and utilized similar techniques (Isager and Skydsgaard 1992: 108–14). For example, although the studies of Jameson (1977), Millett (1986), and Osborne (1987) all acknowledge that most of the source material for Greek farming relates to wealthy farmers, all vacillate between information relating to large and small farmers with little detailed consideration as to how the scale of operations (and concomitant social position) might affect strategies and technologies.[6] Burford (1993: 83–4) argued that all Greek farmers, rich and poor, shared 'peasant-like' attitudes. Similarly, information from the Roman agricultural writers has often been applied to Greek agricultural enterprises, as for example by Amouretti (1986). Although she recognizes the strong relationship between technology on the one hand and social and economic circumstances on the other (Amouretti 1986: 259), she is nonetheless prepared to use the Roman agronomic treatises explicitly to 'fill in the blanks' in the Greek sources in her descriptions of cultivation techniques (Amouretti 1986: 11). Here I shall attempt to characterize larger-scale Greek farms, and to place them in their socio-economic settings before moving on to discuss the techniques of olive cultivation within them.

This is not to say that there were no poorer farmers. The literary evidence suggests that there were many, but there is little secure or precise evidence for them and their activities. It seems likely that there were many different kinds of farmers ranging broadly across the socio-economic spectrum, and that the social order of Greek poleis was not sharply divided into simple groups of 'rich' and 'poor,' or 'large estates' and 'small estates'. However, the fragmentary nature of the sources means that we cannot determine with any accuracy the composition of this spectrum. The bias of most of our source material towards its wealthy end often forces modern academic discourse into such artificial dichotomies. When I have used such crude distinctions in this work I have not forgotten that what can

[6] Isager and Skydsgaard (1992) ignore the problem altogether, and follow Burford's (1977/8) idea that 'farms' in Attica were of approximately equal sizes.

only be seen as broad banding now was probably once rainbow shading, now faded and lost to us.

2.2. WEALTHY MEN: THE ECONOMY IN THE HOUSEHOLD

2.2.1. Large and Small Households

In classical Greece most people most of the time did not conceptually leave their own households to 'go to work', even if they left the physical confines of the house. Fields, workshops, slaves, and buildings, regardless of their physical locations, were all part of the household in conceptual terms.[7] For classical Greece, then, 'the economy' might be defined as the aggregate of economic activities of households and the relationships (both vertical and horizontal) between households: it corresponds to and is entwined with the political system, but is not identical to it. Inevitably, households' economic activities must normally have been acted out within the social and political fabric of polis life. However, crucially to this work, if most of the agricultural land within the territory of any polis was privately owned, then decisions on how to work it will have been taken by households. Hence it is families and households which will have shaped the agrarian landscapes of Greek cities.

The economic 'mentality' of large and small households was not the same, though there were probably shared elements in outlook. However, the primary aims of small-scale farmers were likely to have been be self-sufficiency and survival (Forbes 1976; 1989). Their paramount concern in formulating economic (including agrarian) strategies was thus security (Gallant 1991; Forbes 1982; Scott 1976). For rich households, the aims were fundamentally different. Certainly *autarkia* was important, at least as a moral value and perhaps as an economic value as well. But self-sufficiency is not the same as subsistence for an Alkibiades or an Ischomachos. Their demands and desires for commodities were far greater than mere subsistence level.

[7] Xen., *Oec.* 1.5, defined an *oikos* as consisting of all the property it possesses, even if it is located in different cities; see also Foxhall 1989: 22–5.

2.2.2. Subsistence and Domestic Production

I suggest that large-scale, wealthy, households practised what I shall call *domestic production* rather than *subsistence production.* This entailed the production of surpluses well beyond subsistence in order to maintain, or better yet, enhance and increase, the position, status, and rank of the household and the individuals within it. Each household was competing with other households practising these strategies, while simultaneously being economically, politically, and socially entangled with its competitors. In such a milieu, no household could afford just to stop when it thought it had 'enough'.[8] On the contrary, domestic needs for surpluses could be fabulously high, and the constant stream of sumptuary legislation in classical antiquity bears witness to the fact that it was not lack of desire for *household* growth and acquisition which limited production. The large scale of wealthy households and their high domestic (as opposed to subsistence) requirements result in quite different economic aims and strategies. If security and avoidance of risk characterize 'peasant' households, opportunism must be the chief characteristic of wealthy households, a theme to be developed further below.

For wealthy Greeks, 'household' provided not only the fundamental institutions for social organization but what one would call in the modern world the 'entrepreneurial' and 'managerial' frameworks of economic organization and decision making as well.[9] It was the aggregate of productive units within households, organized and directed within the household, which made up 'the economy' of the city, region, or, ultimately, even larger units. Athens never did

[8] Scholars have regularly conflated subsistence production with what I call domestic production because of the assumption that large, wealthy, households had aims similar or identical to 'peasant' households, Finley 1983a: 72, 1999: 138; Millett 1984; Sahlins 1974.

[9] I am well aware that 'managerial' and 'entrepreneurial' are anachronistic terms, and I use them here to stand for concepts which are quite different from those they signify in the modern world of business and economics, hence the scare quotes. As noted in section 2.1.2, there is no easy solution to the problem of anachronistic and thereby potentially misleading terminology. The idea that these terms convey is that households served as the institutional framework within which production was initiated, executed, and organized.

take Xenophon's (*Poroi* 4.13–21) advice and buy large numbers of slaves to work the silver mines directly, more or less as a 'corporation'. Instead, the complex of productive activities which occurred in the silver mines of south Attica were managerially and conceptually fragmented into smaller units of varying sizes and configurations, belonging to different households, and organized from within these individual households as separate units within the total enterprise of each. Sometimes this might mean a unit as small as a single slave (Andoc. 1.38), sometimes a whole ore washery (Dem. 37)—an installation for crushing silver ore and using running water to separate the metal-bearing component preliminary to smelting. The point is not that Athenians were incapable of forming corporations, or, as Finley argued (1999: 144), and others still accept (Miekle 2002), that the absence of corporations reflects the absence of a need to pool capital resources because 'the prevailing mentality was acquisitive but not productive' (Finley 1984: 44). Xenophon's plan was never taken up because the institutional structures of Greek society were alien to the idea of the corporation, being firmly rooted in alternative institutions, the social relationships of the household. This meant that any 'productive mentality' which might have existed would have been limited to functioning in small, inward-looking, and more constrained contexts than the modern idea of the corporation.[10] To work the silver mines, resources were in many senses pooled, and mining was undoubtedly productive.

For the purposes of this book, the area of economic activity delimited by the concept of polis is generally most relevant. Although individual wealthy Athenians were sometimes able to exploit land outside Attica (see Tables 2.1 and 2.2, and section 2.2.3), this was an anomalous by-product of Athenian imperial activities. What is interesting is that the scale and ambitions of the household of an Adeimantos, or even an Alkibiades, never aspire to those of a Pliny. In part this may be the result of the risks associated with managing and exploiting resources ultimately under the political control of another polis: the integrated and centralized bureaucracy

[10] In contrast, one of the most extraordinary features of the Roman world was the capacity for Roman society to transform and expand the institution of the household (e.g. the Imperial household) to operate (and thus to produce) on a much larger organizational scale.

Table 2.1. Adeimantos son of Leukolophides of Skambonides: surviving possessions in the Attic Stelae.

IG I³ **422**
187–904 shadufs and a large trough on land in Xypetnaion

182–6l and (specifications and location lost)
178–811 and (specifications and location lost)

IG I³ **426** [skilled slaves and equipment—prices missing]
10–39 Phrygian man
a man, Apollophanes
Charias, *obeliskopoios* [spit or nail maker]
Aristarkhos, *skutotomos* [leather worker]
his equipment: small table, 2 couches, table, sleeping pallets, building timber, and 8 unpreserved and unidentified items.
Saturos, *skutotomos* [leather worker]
[3 lines missing and 3 lines that seem to have been equipment]

44–51 [Thasian farm specialising in vines]
44 man, Aristomachos [bailiff?]
45–6l and and *oikia* in Thasos in I--
large numbers of good and bad *pithoi* with lids
590(?) amphorai of wine (capacity: 3 choai) = 8.64 litres each = 5098 litres wine total.

106–7 income from rents on land that had been owned by Adeimantos (cf. line 100): 1632 dr., 4 ob. [if a rent of around 8% of the capital value is assumed, this makes for a capital value of about 3 T, 2408 dr].

142 something unidentifiable worth 520+ dr.

IG I³ **430**
1–4 'oakery' and 'pinery' and oikia in B–, 8 pithoi in the *oikia,* and Kudimachos, slave of Adeimantos [who presumably managed the 'oakery' and 'pinery'].

10–12 harvested crops [cereals or other arable?], worth 50 dr., from land in Ophruneion.
27–8 sale of slave, Saturos, 170 dr.

of the Roman Empire eliminated this drawback to a considerable extent. It seems that more generally the territorial limits of poleis and the exclusivity of land ownership, normally limited to citizens, regularly bounded the agrarian activities even of the wealthiest citizens.

Agriculture was without doubt the primary and predominant productive activity in the classical world. As shown in more detail below, many 'productive' activities (and productive 'capital investment' in them) were engaged in by households: these were, after all, the most direct (though not the only route) to the acquisition of wealth which maintained the social and political standing of the household. However, because these productive economic activities are contained and constrained by the framework of household structures, it is easy to miss them if you look in the wrong place:

Table 2.2. Axiochos son of Alkibiades of Skambonides: surviving possessions on the Attic Stelae.

IG I³ 422
194–204 [slaves]
Arete, Thracian woman (361 dr., for all 3?)
Grulion, Thracian man
Habrosune, Thracian woman

Dionusios, Scythian bronze smith (155 dr.)

income from rents on fields (*choria*) in Tho-- which had been owned by Axiochos, 150 dr. [if a rent of around 8% of the capital value is assumed, this makes for a capital value of 1875 dr.]

IG I³ 424
10–16 apartment house
total of houses [*oikiai*]—large sum of money not preserved.
foreign agricultural land—details not preserved

IG I³ 426
101–2, income from rents on land owned by Axiochos,
108–1111633 dr. 2.5 ob. [if a rent of around 8% of the capital value is assumed, this makes for a capital value of about 3 T, 2417 dr.]

item not preserved, more rents? 250 dr.

item not preserved, more rents? 162 dr, 4 ob.

IG I³ 427
52–85 [equipment and fittings from a country house]
5 phidaknai [small pithoi]: 9 dr; 11 dr.; 4 dr., 4ob.; 4 dr., 3 ob; 4 dr.
funnel [no price, goes with next item?]
lead pipe 2 dr., 2 ob.
written board/picture 60 dr.
another small one 6 dr., 4 ob.
painted(?) picture 5+ dr.

land which had belonged to Axiochos . . . [further details missing]
[poorly preserved entry] 2040 dr. (?)
[poorly preserved entry] 1590 dr. (?)

area of land (in plethra) with *oikia*, another to the metics/merchants [no price]

3 plethra arable land with vines 1900 dr. [goes with last item?]
oikia in the countryside [*agroi*]
another piece of arable land, with olives(?), 3 plethra 6100 dr.
[something unidentifiable] with vines; [something unidentifiable] in Abydos 310 dr
[something unidentifiable] in Klazomenai 200 dr.

IG I³ 430
6–7 a man, Olas 195 dr.
8–9 Messenian man 130 dr.
24–5 Keph--, slave 195 dr.
33–5 crops in the field(?) 20 dr.

beyond the household they cannot be found. This also means that 'investment' remained on a small scale even in the largest and wealthiest households, limited by the resources and labour available within any individual household.

Now this is not to say with Finley (1983a: 69) that '"firm" and private household were one and the same'. That is to describe the organizational and conceptual structures of 'household' too simply, and, even more misleadingly, to 'primitivize' wealthy Greeks. For some purposes, and certainly at the level of ideology, business wealth and personal wealth were inseparable, as was for that matter the wealth of different individuals within a household (Foxhall 1989). But at another level, in many contexts, individual economic enterprises managed by a household seem generally to have been separated into discrete units.

The accumulation of self-contained units within a large household, combined with the lack of 'managerial' specialization also provided a means by which wealthy Greeks could engage indirectly in profitable, but socially undesirable, activities. Sometimes this might be used to conceal more direct involvement in such activities. Andokides (1.137) refers in a roundabout way to his shipping and trade concerns, and he admits to more personal involvement than most wealthy Athenians.

2.2.3. Economic Units within Households: Diversity and Domestic Production

The households of rich Athenians are characterized by their extraordinary diversity of activities (Millett 1991: 165–9).[11] A great many examples could be cited, despite the fact that it is difficult to tease complete household property holdings from the sources: too often it was in someone's interest to exaggerate the wealth of another or to underplay their own. For modern scholars, this diversity has often

[11] My guess is that the same applied in other Greek cities as well, but the evidence outside Athens is limited (Davies 1981). Although diversity may well have characterized the economic activities of smaller households as well, both level and scale of 'enterprises' and the range of diversity is likely to have been less, simply because they were poorer.

typified the irrationality (in the terms of formal economic theory) of Greek attitudes to profit and the acquisition of wealth, which was simply 'unorganized and unsystematic' (Millett 1991: 166). One might better ask with Cartledge (2002: 21–2) 'whose rationality?' I would argue that, on the contrary, this diversity is part of a quasi-deliberate strategy for generally maximizing potential opportunities for profit and the acquisition of wealth within the context of volatile friendships, political alliances, and, in the broadest sense, environments which the primary actors may have generally perceived as being unstable.

Demosthenes' estate consisted of workshops with thirty two slaves making swords and twenty two or so slaves making couches; supplies of raw materials for both; a house with domestic slaves, furniture, jewellery and clothing (much of this probably produced by the domestic slaves and other women of the household); bottomry loans; interest-bearing loans; *eranos* loans; and money in three different banking establishments (essentially more outstanding loans). Similar in configuration, but on a smaller scale, is the estate of Komon (Dem. 48. 12–13) which consisted of a workshop of slave *sakchuphantai* (sail makers or sackcloth weavers) in or attached to his dwelling house, and another, smaller house which was a workshop of *pharmakotribai* (drug or colour grinders). Unusually, there is no agricultural land with either of these estates. Andokides (1) possessed merchant ships, a large and valuable town house in Athens, agricultural land in Attica, and agricultural land in Cyprus, and probably much more as well. The speaker of Demosthenes 47 had a flock of fifty fine-wooled sheep with a slave shepherd (a cash generating enterprise in their own right), a farmhouse with tower, domestic slaves, furniture and equipment, possibly other agricultural slaves, possibly a house in the Piraeus, and probably other agricultural land. Two small buildings, sixty plethra (5.4 ha) of plains land, and a bath house allegedly represent only a small proportion of Dikaiogenes' estate (Is. 5.23–4). The speaker of Isaios 9.44 claims to own only land worth 5000 drachmai at Oinoe, land worth 3000 drachmai at Prospalta, and a house in Athens worth 2000 drachmai, though he is probably under-declaring. Aischines (1.97) claims that Timarchos inherited from his father a house south of the Akropolis; an *eschatia* at Sphettos; (agricultural?) land at Alopeke; a leather workshop with nine or ten skilled slaves and a slave manager; some kind of specialist textile workshop with at least two skilled slaves, one male and one female, operating it; along with outstanding (probably interest-bearing) loans and household furnishings. For these proprietors, of course, this is all that we know about: they may have owned much more.[12]

The fragmented and fragmentary estates of Axiochos son of Alkibiades and Adeimantos son of Leukolophides, in so far as they can be teased out of the Attic Stelae (*IG* I³ 421–430), provide interesting examples of the same sort of diversity in even greater detail.[13] Details of the agricultural enterprises arch discussed in detail (see Tables 2.1 and 2.2 and section 3.2). But in terms of the overall range of activities, Axiochos owned agricultural land and equipment in a number of places, various houses and buildings including an apartment house (*sunoikion*), domestic slaves, and a bronzesmith (and probably his workshop and equipment). Adeimantos owned what seems to have been an irrigated garden with four water-lifting devices, forest land ('oakery' and 'pinery') potentially exploitable for timber and other forest products, a farm in Thasos specialising in vines, including its slave-bailiff and around 6000 litres of wine in store (these last two must be specifically cash generating enterprizes), various other plots of agricultural land, an *obeliskopoios* (spit or nail maker) and probably his workshop and equipment, two *skutotomoi* (leather workers) and their equipment, and various domestic slaves. 'Diverse' is certainly no exaggeration here.

It should be obvious from these examples that many of the economic enterprises (including agricultural ones) within these large households were 'self-contained' units, easily detachable by sale, gift, or dowry, but were managed under the organizational umbrella of the household. A rather amusing example of this comes in Hypereides 3 where the speaker has purchased not only the love of his life (a slave boy working in a perfumer's shop) along with his father, the shop, and its equipment, but also a number of bad debts.

[12] See also the properties of Theophon and Sosikles, Isaios 9, in Chapter 3.

[13] Osborne (1987: 21–2) assumed that the defendants in the cases of the profanation of the mysteries and the mutilation of the herms were all young men. In fact Davies (1971: 16–7) showed conclusively that Axiochos was fully mature, probably in his fifties in 415 BCE.

Note that these are debts of this specific enterprise, apparently separable from the rest of the business and household affairs of the former owner of the shop. More mundanely, in Isaios 9.41–3 one of the four daughters of Stratokles inherited the patrimonial estate of her maternal uncle. Although it was managed and worked by her father for nine years, and the income from it was used to improve the economic standing of her whole natal household (Is. 9.45), the property was nonetheless 'detachable' from the rest of the household's property, and she apparently received it on marriage. Similarly, the speaker of Isaios 9.28–9 claimed that his father managed and improved by investment the land inherited by his half-brother, and that it was passed on to him worth more than when he had received it.

It should also be noted that when property was divided on inheritance, individual enterprises usually appear to have remained intact as far as possible. In Demosthenes 48 the two lots consisted of one workshop each. Similarly, when the estate of Pasion (Dem 36; 45; 46), a metic banker of slave origin, was divided at the request of the guardians of his younger son Pasikles, the elder son, Apollodoros, was offered the choice of his father's bank or his shield-making workshop (Dem. 36.8–11) and chose the latter. The property of this family, however, has a strange and perhaps unusual history, as a result of the 'slave to citizen' and 'rags to riches' transformation of Pasion and Phormion, Pasion's ex-slave who married his widow (Millett 1991; Cohen 1992).

2.2.4. Sources of Diversity and Opportunism

How did this extraordinary diversity in household enterprises come about? The answer is simple on one level: it is a logical outcome of a society where the key unit of social organization is the household, and the 'management' of productive and economic activities is contained within its framework. On another level, this diversity emanates from other elements of Athenian (and more widely Greek) political and social organization. This exercise of definition itself illuminates the way in which economic activities were organized and run within households.

Fundamental among these processes is inheritance (Burford 1993: 33–48). Ancient Greeks, including Athenians, used a system of partible inheritance. Heirs (direct lineal descendants, in the first instance) received equal shares of the property, though generally (as in Athens), female shares were smaller than those apportioned to males and were handed over, all or in part, as dowry at the time of marriage (Foxhall 1989 and see section 2.4.4). As noted previously, when household properties were split up on inheritance, generally discrete 'enterprises', including agricultural ones, were kept intact, as far it was possible to do so and still divide the property fairly. Obviously such a strategy would be easier to operate in wealthy families than in poorer ones. But householders must have always been well aware that the day of reckoning, when they would have to pass things on to their children, would come eventually (Pl. *Resp.* 330b). This must have been an added incentive to keeping economic activities in small, discrete units, since it would help reduce the chances of quarrels when the property was divided. In any case, the limits of legal sanctions and structures, the limits of 'managerial' and 'administrative' technologies (for example, rudimentary accounting procedures), and the likelihood that few people outside the household were believed to be trustworthy (see section 2 2.6) probably also encouraged smaller individual economic units. Hence, accidents of inheritance, not only from the natal household, but also from households of collateral kin who died without direct lineal descendants, or via *epikleroi*, or even occasionally via dowry,[14] might contribute to the diversity of a rich household's economic 'enterprises', and some of the diversity of 'enterprises' received ultimately by a household might be quite unexpected.

Spare cash was often loaned out, either as interest-bearing loans with a direct economic return expectable, or as interest-free loans (*eranos* loans), with an economic return equally expected, but not necessarily directly or immediately (Millett 1991 and see section 2.2.6). Both kinds of loans were normally secured by property. If the recipient defaulted, the lender received the property, or a share of its sale price

[14] See, for example, a house (*oikia*) in the Athenian Agora given to a woman as dowry (Finley 1951: 192, no. 175A). Cf. the practice of metics using citizen agents in loans secured by land (Millett 1991: 224–9).

to the value of the loan as a substitute for his money. As a result of the precarious financial wheelings and dealings in which many heads of household seem to have engaged, it appears not to have been unusual for creditors to end up with unexpected pieces of real property or other economic enterprises. For example, the *klinopoioi* belonging to Demosthenes' father came to him when someone defaulted on a loan of 40 minas he had made (Dem. 27.9). He probably never intended to end up in charge of a couch-making workshop, and it is clear that the character of property obtained in this way was almost entirely at the whim of chance. I do not believe this can be explained simply as chaotic or unsophisticated management; rather it was intrinsic to the system of obtaining and managing resources. Wealthy Greek householders appear to have revelled in these situations and manipulated them to their own advantage whenever such chances arose: this was Tyche at her best. Property received as security on defaulted loans must therefore have contributed to the highly variegated composition of rich households' range of economic activities.

Opportunities to acquire wealth-generating resources were seized as and when they appeared. These rich householders aimed to maximize the opportunities for accumulating wealth over the long term. The lack of potential for investment (Miekle 2002; Millett 1991: 169) is in a sense the logical outcome of fragmented and atomized small-scale economic 'units' contained within households. However, the other side of the coin, as it were, is that the social and political contexts of Greek cities permitted wealthy men (and occasionally women) to use the consumption of wealth to create representations of themselves and to engage in transactions through which they could enhance their own and their family's status. This results in quite different strategies than those associated with more familiar aim in our society and economy of maximizing the profits of specific, individual enterprises in the comparatively short-term. In any period or place without the legal expedient of the limited liability company *vel sim.*, and in the absence of state enforcement mechanisms (e.g. in cases of debt, default, etc.), the failure of economic 'enterprises' run within a household must have jeopardized that household overall, even if such 'enterprises' often consisted of self-contained units. Diversity must have limited the risks of failure when disaster struck. Given the relatively short life expectancies and the problem of endemic disease in cities like ancient

Athens, the risk of skilled slaves dying, especially those who had come from elsewhere, must have been considerable.[15]

2.2.5. Cash, Opportunism and Diversity: the Problem of Timescale for Using Resources

When rich households had spare cash (and at times they seem to have had large amounts), it was stored like any other commodity in chests and/or jars. Loans, including *eranos* loans, were a way of doing something with cash. Putting cash in a 'bank' may simply have meant that someone else was conducting the actual business of loaning out the money (Dem. 27.11), though it is possible that such accounts were rare and set up for special purposes (Millett 1991: 215). Cohen (1992: 111–20, 190–4) suggests that the use of banks was more widespread, but even if this is correct it seems likely that only a wealthy minority used them regularly.

It has been claimed (Millett 1991: 64–74; Miekle 2002: 243) that loans, including *eranos* loans, were purely social, used for non-productive purposes (e.g. performing a trierarchy or other public office) and conspicuous consumption (e.g. dowries).[16] Millett (1991: 17–18, 71–2) points out that the roles of lending, borrowing, and credit in ancient Athens share more in common with modern Third World systems than with the world of Adam Smith. One might then question whether 'productive' and 'non-productive', in the terms of modern economic theory, are particularly helpful categories.

Economic behaviour in classical Greece, most clearly visible in the Athenian sources, operated on different timescales than those to which modern economic thought is most accustomed. This idea could be related to Parry and Bloch's (1989) notions of long- and short-term transactional orders as applied by Kurke (1999) and von

[15] Cf. Landers 1987, Landers and Mouzas 1988, on the high mortality rate of the population of London, especially immigrants from rural areas, in the sixteenth through to the ninteenth centuries from endemic and epidemic disease.

[16] Millett 1991 has used the terms 'productive' and 'unproductive' in relation to the borrower's motivation: obviously all interest-bearing loans are 'productive' in the general sense of the word for the lender. As Millett (1991: 230) has carefully defined the distinction: 'productive credit may be defined more closely as referring to loans taken out with the expectation that the resulting return will be greater than the interest charge'.

Reden (1995), or Bourdieu's notion of the simultaneous operation of different 'calendars' of the habitus, linked by common symbolic elements (1977: 96–158), but goes beyond these. Wealthy householders appear to have thought on a different, longer-term time scale than we are used to in terms of the practicalities of organizing and running their productive and economic wealth-generating activities. What might appear to be a 'non-productive' loan, offering only symbolic benefits (Millett 1991: 71–2, 229–32), can in fact be shown to operate on a different timescale as well as perhaps in a different transactional order, ultimately also bringing material gains.

Virtually all the uses to which 'non-productive' loans were put in classical Athens can be shown to be capable of producing straightforward material gain on medium- to long-term timescales. Frequently, loans were used for performing liturgies, or for the expenses of public office. Undoubtedly there were symbolic and political aspects, but prominent public positions carried potential profit (Millett 1991: 230). Greeks from Hesiod (*WD* 40–5) onward, and certainly Athenians, expected that office-holders would make profits from their posts, either illicitly, or in a legal but distinctly dubious manner (Harvey 1985; Millett 1991: 85). Certainly opponents at law in Athens were always ready either to accuse opponents of making profits via office-holding at the public expense (Aesch. 2.3, 23, 71, 79) or else bragging that they themselves had held office or performed liturgies without dipping their fingers in the public till or taking bribes. As in our own world, political connections were potentially profitable in the long-term: note, for instance, the eagerness of rich Athenian politicians to serve as foreign ambassadors or *proxenoi*. Office-holding, then, undoubtedly had an economic facet, but the economic return was not necessarily immediate or visible.

Similarly, borrowing for dowries is frequently characterized as 'unproductive' (Millett 1991: 62–3). In the terms of modern economic theory, of course it is. But is that the most useful way to analyse such borrowing? Just as landholding and property served to validate the status of male citizens, dowries in Athens validated the citizen status and legitimate matrimonial status of a woman. In the law courts payment of a dowry was regularly brought forward as evidence to show that a woman was legitimately married (Foxhall 1989). If dowry served to some extent as female inheritance (a girl's share of the property of her natal household) but had to be delivered at the time of the girl's marriage (Foxhall 1989: 32–3), the moment of delivery is likely to have been a particularly inconvenient one in the lifecycle of the natal household (Gallant 1991: 11–33). Parents, still at the prime of life, needed to keep hold of their economic resources, especially capital resources, both to maintain themselves and to support other children still at home. Since girls generally married very young (Cox 1998: 142), the natal household might expect to be economically active for many years.

The social institution of the *eranos* loan got round the problem very neatly. Instead, the father of the bride raised an interest-free loan from among his friends, but retained his 'capital' resources, the income from which not only paid back the loan, but generated further income for many years longer, which could ultimately increase the estate to be passed on to his heirs. This may have provided part of the impetus among the wealthy at least to give dowries in the form of cash or outstanding, interest-bearing loans rather than real property.

The wealthy elite of Athens used the *eranos* system as a means of exclusive self-generation. The rich could maintain their position of status, wealth, and power without bankrupting themselves or letting in others. This is not to say that only the rich used the *eranos* system: we simply do not have sufficient evidence to know if and how it might have been used by other classes.

Eranos loans, stored commodities and raw materials, and agricultural produce, are oddly comparable. All bear returns only in the medium- to long-term. Time scale was vital to the use of resources for ancient households. This is especially so of the economy largely located within the 'entrepreneurial' and 'management' structures of the household. Households were intended to continue into the future, if not indefinitely, and at least for more than a single generation (Foxhall 1989: 28–9 and n.32; 1994: 139; cf. Plato *Resp.* 330b). The use of time scales which think one or more generations ahead have generally not been a feature of modern economic thinking. Households acquired resources, partly, in order to pass them on to the next generation. It was of course, perfectly 'rational' in the ordinary sense of the word, given the aims of Greek households, to consider the long-term, if a household could afford to do so.

2.2.6. The Household and the Limits of Trust

As the household provided the framework for economic activities, it also furnished the limits of trust (Gallant 1991: 152–3). Friendship even with kin outside the household was volatile and liable to betrayal, whether perceived or real. The cases of Demosthenes' battles against his guardians (Dem. 27–30), or Andokides' opponents in his trial for impiety (Andoc. 1) spring immediately to mind, but many other cases could be cited from the Attic orators. Similarly, in many modern Mediterranean societies where the household is the fundamental social and economic unit, it functions similarly as a delimiter of trust (DuBoulay 1971: ch. 2, Pina-Cabral 1986: ch. 1, Pitt-Rivers 1971). Ian Morris's (2000: 116) observation that *philia* is part of the ideology of the community of Athenian citizens is significant here; the problematic discourse about trust which goes alongside it suggests this is an ideal which was not often attained.

That trust outside the household was a problem in economic ventures is clear in the speeches of the Attic orators. For example, in Demosthenes 37.11, the speaker claims that when he found Panaitios, a man with whom he had become engaged in business dealings, 'doing the dirty' on him, he went and complained to Mnesikles, the man who had originally introduced and recommended Panaitios to him. It is noteworthy, too, that even *eranos* loans to 'friends' were normally secured (Millett 1991: 153–4).

The contrast is to be found in Demosthenes 48.14–15, where the speaker alleges that Komon, the deceased owner of the property which is being divided, had been cheated by one of his slaves, 'whom Komon thought was especially faithful (*piston*) to him . . . this slave had a good understanding of nearly all Komon's other affairs' (cf. Xen. *Oec.* 9.11–12; 12.5). Although rich Athenians were surely not so naïve that they believed all slaves were trustworthy, and clearly householders of free status exerted control over slaves which they would not have had over free kin or friends, why did they trust them with important matters at all (Hunter 1994: 181–4)? The answer must be that they were part of the household in a way that even non-resident kin were not, and therefore within the conceptual limits of trust. It is in this light that we must see Sokrates' complaint that men lavish more care and affection on their slaves than on their friends (Xen. *Mem.* 2.4), and Theophrastos' parody of the 'Rustic' as a man who 'does not trust his friends and relatives, but consults his slaves about important matters' (Theophr. *Char.* 4.2). This must also, incidentally, be one of the striking advantages of using a slave labour force for Greek proprietors, especially as overseers. Because they were part of the household, they were firmly within the limits of trust, as well as within the limits of the householder's control, and avoided the negative social valuation attached to wage labour in Greek culture (Xen. *Mem.* 2.7).

2.3. RISK AND INFORMATION

Rich Athenians (and, perhaps, most wealthy Greeks) were opportunistic in their approach to acquiring economic 'enterprises' and resources. Risk must certainly have been considered in so far as excessive risk-taking might jeopardize the whole household. However, risks were most certainly taken in hopes of great profits by the rich who could afford to do so: the trade in high interest bottomry loans well documented by the orators provides an example (Millett 1983). But no one in his right mind would have invested the whole of his household's resources in this way. The greatest perceived threat, however, was probably that friends (even kin) would let one down (Foxhall 1998b). Hence, when opportunities arose in whatever way, to acquire resources (including agricultural ones) which were reasonably safe, these chances were seized.

The resulting ragbag of activities looks haphazard from a modern point of view. But the very eclecticism of economic enterprises organized within households may explain the lack of specialization at 'management' level. There is abundant evidence of craft specialization in ancient Greece, in particular from Athens (Harris 2002). However, nearly all is documented at levels within household structures. A number of inscriptions list people living in Athens (largely foreigners, metics, and slaves) with their occupations (Harris 2002; *IG* II^2 1553 ff.; *IG* II^2 10 [Tod 1948: 8–13; Rhodes and Osborne 2003: 20–6, no. 4]). Such occupations as figurine maker, walnut seller, tub maker, baker, and so forth, testify to the specialized division of labour in fourth century BCE Athens. However, given a system in which the household provided the organizational framework for economic activities, technical specialization would have been positively disadvantageous for opportunistic householders themselves, since they could never predict what enterprises they might acquire over a lifetime. Over-commitment in one direction would have cut down flexibility. Instead, householders specialized in the building of economic networks set within socio-political relations, and the accumulation and networking of information. Rich Athenian proprietors spent a considerable amount of their time making contacts and gathering information. They were not locked into rigid state bureaucracies or pre-set career patterns, much less nine-to-five jobs. The references to this class of men hanging about the Agora and other public places, meeting at each others' houses, whether at *symposia* or other gatherings, are innumerable. Public life was economic life as much as it was political and social life. Though the philosophers may stress the civic concerns of public life and the primacy of the state (Pl. *Resp.* 368e–369), the accounts from Athenian courts suggest that atomic economic concerns took priority over concern for the city's welfare most of the time. By meeting many people, and listening to the stories told about them by others, by becoming known and making many 'friends' and contacts, wealthy Greeks became aware of opportunities by which they might enhance their households, as well as of hazards which might threaten them. If Greek householders could ever be considered 'specialized' in any sense, it may have been as gatherers and evaluators of information. At one level, such activity is no more and no less 'unproductive' than modern television advertising or business entertaining. Though such activities are not directly or quantifiably profitable, they are felt by those engaging in them to lead usefully to significant benefits by creating information networks and spreading knowledge.

2.4. CONCLUSION: TOWN AND COUNTRY?

The concept of the household as the context of economic organization, and the concept of the household as the definition of the limits of trust are inextricably interlinked. Wealthy households managing a diversity of small, 'self-contained' enterprises, acquired opportunistically, were intrinsic to the political and social organization of classical Athens. This is different from the way in which post-medieval economies (including our own) are 'embedded' in culture. Within their own terms, the activities of these wealthy households were certainly 'economic'. But, timescales, the parameters of risk and hazard, and most especially the organizational structures and their location within households, combine to create an alternative rationality that is not the same as ours. Beyond this, the whole strategy of domestic production was geared to a kind of consumption, especially of special, 'semi-luxury' commodities, which itself played a major role in constructing and validating the status of these consumers.

Although households were conceptually unified they were not necessarily spatially coherent: the activities and resources of a single household might be spread all over Attica and Athens, and beyond, crossing the boundaries of urban and rural. Rich Athenians owned both country and town properties, some of which were dwelling houses, and some of which were workshops or agricultural holdings. We can see from sites like Olynthos that both elements could be combined in a single structure (Cahill 2002; see Chapter 6). As Osborne (1987) pointed out long ago, town and country were interdependent. Who else but the craftsmen who carved stone in the city for urban buildings would have had the skills to make the stone troughs, olive mills and pressbeds of the countryside? Consumers and producers did not correspond to a divide between town and the country. Instead, the dichotomy was between those who were truly autonomous '*oikonomoi*' household managers of wealthy households, and those who were compelled to labour within them.

3

The Agricultural Holdings of Large-Scale Households

3.1. THE IMPORTANCE OF LAND

Land was not simply one kind of resource among others. For most prosperous Greek households, land must have been the most important element among the selection of economic activities in which they were engaged. The political and symbolic significance of land was as important as its economic value. Land ownership was the prerogative of the citizen, and thus distinguished the enfranchised members of a polis from non-members. The ownership of land had both civic and social implications for ranking and status within the polis (Foxhall 1989 40–3; 2002; Burford 1993: 2). But that is not to say that land ownership was esteemed purely because of its symbolic importance.

In predominantly agrarian societies like those of ancient Greece, land was certainly the most secure resource precisely because land ownership provided direct control of the means of production (Scott 1987). Rich Athenians had little sentimental attachment to particular pieces of real property, except for relatively small plots with family tombs (Humphreys 1980: 97–8, 105–112; Burford 1993: 97–8), and perhaps occasionally, 'ancestral' houses (Andoc. 1.146; Is. 5.11), few (if any) of which were older than a hundred years. The frequency with which land was put up as security for loans (and thereby at risk of being lost if the borrower defaulted on the debt) suggests it was dealt with in the same opportunistic way as other economic resources (Osborne 1985b: 53) Athenians were not fearful of the risks of alienation per se (contra Burford 1977: 167; cf. Burford 1993: 49–55). On the other hand, if the chance to acquire land arose, it was taken

(Osborne 1987: 38; Burford 1993: 49–55; Lambert 1997). The ownership of large amounts of land (within the relative terms of the limited scale of the territory of a Greek city-state) allowed the satisfaction of *domestic* (not simply *subsistence*) needs from within the household. Thus the dependence of the household for 'necessities' on agents outside the limits of trust was reduced: dependence, that is, on other households and unpredictable market transactions. This is the practical significance of the moral value of *autarkia*: the implication of large-scale and direct control by a household of the means of production, which limited the extent to which that household was subject to other households and institutions. In other words, it is not simply 'self-sufficiency'; it also represents that key attribute of the ideal male citizen, 'autonomy'.

Despite the monopoly on land ownership by polis citizens and its consequent symbolic and political significance, money and land were not necessarily two separate spheres of activity, or even necessarily linked to different transactional orders (Finley 1953: 72–3; Millett 1991: 225). Certainly there is no good evidence for the separation of money and land within households (see Chapter 2). In Athens, from where so many of our sources emanate, metics could not own land, and therefore dealt only in money, at least while they were in Athens. But some metics did not necessarily spend the whole of their lives or even all of their time in Athens. Of the metics who settled in Athens voluntarily in the first place (i.e. those who were not freed slaves), some, possibly most, must have owned or had claims to land in other cities. Thus a metic's activities in Athens may in some cases have represented only part of the economic resources of his household (Whitehead 1977: 17–19).

3.2. DIVERSITY WITHIN AGRICULTURAL HOLDINGS

It is generally accepted that Greek landed proprietors normally had scattered and fragmented holdings.[1] This is best known for Attica,

[1] Jameson 1977: 130–1; Osborne 1987: 37–40, 1985b: 60–3; Burford 1993: 110–111, 119; Isager and Skydsgaard 1991: 128. Hanson 1999 appears to be one of the few scholars who does not accept this; see Foxhall 1997: 2002.

though there is limited documentation elsewhere (for Sparta see Hodkinson 1986; 1989: 80–2; 2000). A combination of the process of partible inheritance (discussed in section 2.2.4) and the opportunistic acquisition of land were in large part responsible for this pattern of land tenure. The ownership, and, to a lesser extent the lease, of scattered plots and groups of plots along with the broken landscape and diverse micro-environments typical of much of the Mediterranean region of Greece must have encouraged a preponderance of mixed farming, at least within a household (Forbes 1982: Ch. 11; forthcoming; Osborne 1987: 31–3). This is not to say that particular agricultural 'enterprises' could not be specialized: there is evidence to suggest that some were very specialized. But, within a household, the sum total of the plots cultivated and agrarian activities was more likely than not to be mixed. Almost all farmers must have grown some cereals and usually other arable crops as well—this conclusion is certainly borne out by the mix of crops and farmland documented on the Attic Stelae. My guess is that generally the larger and wealthier the estate, the greater the overall diversity was likely to be, and simultaneously the more specialized individual farming units or activities within a household were likely to be.

A glance at the agricultural holdings of wealthy Athenians confirms this picture. The amazing geographical spread and variety of the agricultural holdings of Adeimantos son of Leukolophides and Axiochos son of Alkibiades as listed in the Attic Stelae are outlined above (section 2.2.3 and Tables 2.1 and 2.2). The estate of Theophron at the time of his death (Isaios 9.41) consisted of land at Eleusis worth two talents (12,000 drachmai), sixty sheep, one hundred goats, agricultural equipment and tools, a cavalry horse,[2] and other unspecified property (probably furniture and personal effects). Stratokles left on his death (besides his town house at Melite and his country house at Eleusis, 4000 drachmai in interest-bearing loans, 1000 drachmai in *eranos* loans and 900 drachmai in cash) land at Thria

[2] Not, of course, strictly speaking agricultural, but for military use, personal transport, as well as conspicuous consumption and enhancement of status more generally. Ownership of such an animal implies the possession of agricultural land for its maintenance. A 'bottom of the line' horse cost around 300 drachmai in fourth century BCE Athens, according to Is. 5.43. If this one is as good as the speaker says, it should be worth more than this.

worth 2.5 talents (15,000 drachmai) and 4900 drachmai-worth of agricultural tools and equipment, sheep, barley, wine, and fruits (*oporai*) (Is. 9.42–3). The stored agricultural produce strongly suggests a mixed farming regime.

3.3. AGRICULTURAL 'CAPITAL', 'INVESTMENT' AND THE PROBLEM OF TIMESCALE

Whether we can appropriately talk about 'capital' in relation to ancient Greek farming is a difficult question. Miekle (2002: 242), arguing along lines similar to Finley (1999: 147–7; 1965; 1953: 71) and Jongman (1988: 25–8, 151–2), claims that capital is not a relevant concept in antiquity because decisions based on 'use-values', the small scale of the units of production, and the lack of a fully developed system of credit, did not allow capital investment in a modern sense. For Miekle, the bulk of production in the Greek world of the classical period was done by free peasant proprietors producing at or near subsistence. The rest was done by a relatively small number of craftsmen producing in workshops of very restricted scale, and by chattel-slaves mainly on the estates of the propertied class (Miekle 2002: 242).

The truth of the first part of this assertion is questionable, and we have no quantified data from the world of classical Greece to test the relative production and productivity of larger-scale versus smaller-scale landowners. If the ethic that it was desirable to pass on as much (or even more) to your children as you received yourself (Foxhall 1989: 28; Hesiod *WD* 375–80; Plato *Resp.* 330b; cf. Hanson 1999: 146–7), then the incentive to improve agricultural land is clear. Although these are small-scale improvements, set within household structures, their aggregate impact could have been quite significant, to judge from the high levels of classical sherd material found in the Greek countryside (Pettegrew 2001; Bintliff *et al.* 2002).

It is certainly true that Greeks rarely wrote about 'improvement', and they certainly never wrote about 'capital investment' in the way that modern economists might. However, archaeological and other epigraphical evidence make clear that Greek farmers regularly increased the utility of their land, and thereby its productivity and value, with installations, equipment, and other improvements, as well as by the planting of perennial tree crops.

The technology of investment in improving land is environmentally and regionally specific. In the broken landscapes of the Greek world of the Mediterranean region, with relatively light, shallow soils, and erratic, seasonal rainfall, the use of modern agricultural machinery or 'more advanced' equipment may not represent a significant agricultural improvement.[3] 'Investment' in land in terms of labour intensification or 'capital' inputs are not necessarily so obvious or visible. Also, the problem of timescale re-emerges. If households improved agricultural land with the livelihood of their children in mind, inputs aimed at the longer-term may not be readily apparent as such from a short-term point of view. Agricultural improvements in ancient Greece are examined here under three broad headings: equipment and building works, other kinds of labour intensification, and tree planting. They will be discussed separately in the sections that follow, but they are, of course, interrelated.

3.4. EQUIPMENT AND STRUCTURES

3.4.1. Agricultural Equipment and Buildings

Many different kinds of improvements to land and agricultural fixtures are epigraphically and archaeologically documented for classical Greece (Osborne 1985a). Survey work has been particularly important in this regard, since truly spectacular numbers of classical period structures scattered throughout the countryside have been revealed by archaeological survey in the past thirty years, and these 'farmsteads' themselves are likely to represent some attempt to

[3] See, e.g. Forbes 1982: 206–16, on why ard ploughs continued to be used into modern times, rather than mouldboard ploughs; in contrast to Jongman 1988: 82–7, who incorrectly assumes that technological improvements can be 'absolute' regardless of the ecological setting. Within the past twenty years deep ploughing and bulldozed terracing have been responsible for serious environmental damage in many parts of Greece and Italy (Grove and Rackham 2001: 363).

improve land, if only in the sense that they enabled the input of additional and more continuous labour. Many installations would have been relatively expensive, demanding sizeable inputs of skilled labour and raw materials. Consequently, their construction must have been beyond the means of many poorer farmers and was perhaps sometimes uneconomic even for wealthier farmers unless levels of production of a specific crop made them worthwhile.

There is a great variety of these installations. Stone paved threshing floors have been found in connection with classical period farmhouses in Euboea (Keller 1989). Public threshing floors are mentioned as an element in a plot of land in Attica sold off to a private buyer in the second half of the fourth century officials in stele 2 of the 'Rationes Centesimarum' inscriptions (Lambert 1997: 52, 229). Earthen threshing floors, more common in many parts of the ancient (and modern) Greek world are less archaeologically detectable, especially in field survey. Brunet has carried out a test excavation on one threshing floor on Delos thought to be ancient. This work has so far been published only in brief, but Brunet (1997: 776) has dated it to the first century BCE on the basis of an Athenian bronze coin discovered at the foundation level of the exterior face of one of the surrounding upright stones, although the stratigraphic relationship of the coin to the threshing floor is not clear from the preliminary publication. Threshing floors are specially mentioned as part of Phainippos's *eschatia* (Dem. 42), presumably because they raised the value of the property.[4] Olive and wine presses, olive crushers, and wine treading floors are both epigraphically and archaeologically documented. These are discussed in more detail in Chapter 6, but it is worth noting here that these seem to have been located out in the fields as well as in farmhouses, villages, and towns. On more marginal land, lime kilns and charcoal burning would also have increased productivity, and these activities are well attested. The Berbati 'farmstead' even housed a classical pottery kiln—unusually clear evidence of craft production in the classical countryside (Hjohlman *et al.* 2005).

[4] The speaker of Demosthenes 42 is attempting to show what an exceedingly valuable property it is, de ste Croix 1966. On threshing floors see also Amouretti 1986: 71, 228.

Unfortunately, prices for major, and one would expect, relatively expensive installations are rarely preserved, even in such helpful sources as the Attic Stelae. In *IG* I³ 422.305–6, 10,200 vine props and what is probably the small and removable stone trough from a wine treading floor were sold together for the relatively modest sum of fifty nine drachmai. On the other hand, an unspecified number of beehives, perhaps complete with bees, went for 260 drachmai (*IG* I³ 426.56–7). By comparison, the upper stone of a saddle quern went for seven drachmai, two obols (*IG* I³ 422.290). None of these smaller pieces of equipment is particularly expensive, but the source material does not allow us to cost the construction of an olive press, a wine treading floor, or a threshing floor.

3.4.2. Agricultural Terraces and Field Walls

On steep land, one important capital improvement particularly relevant to olive cultivation might appear to be the construction of terrace walls. Scholars have regularly postulated the building of terraces as a key element in the expansion of farmers into sloping, marginal lands (*eschatiai*) in periods of economic buoyancy (Runnels and van Andel 1987: 327–30; Jameson *et al.* 1994: 399), or in the face of increasing population (Hanson 1999: 47–50, 65–7). Terraces have also been regularly associated with the cultivation of olives and vines (Hanson 1999:80–2; Lohmann 1993), although, of course, they can be used for planting cereals and many other crops as well (Acheson 1997: 175–6; Grove and Rackham 2001: 110). Bradford (1956; 1957) thought that he could distinguish ancient terracing in Attica from aerial photographs, and Lohman (1993) has also claimed a classical date for terracing in southern Attica (see also Chapter 6.5.4). Doukellis (1994) has attempted to trace ancient terracing and land measurement systems in modern field boundaries. Brunet (1999; Brunet and Poupet 1997) carried out test excavations on terraces on Delos, claiming a Hellenistic (first century BCE) date for them. Most recently Price and Nixon (2005) have helpfully set out the case for the use of terracing in antiquity.

Nonetheless, it remains the case that few agricultural terraces have been identified in the course of either excavation or survey which can be dated with certainty to ancient times (Foxhall 1996). To my knowledge there are none which can be *securely* dated to the Archaic or Classical period. This does not mean they did not exist, but it is certainly the case that the available archaeological and written evidence is insufficiently clear, full or precise to document them securely. I have argued elsewhere (Foxhall 1996 and see below Chapter. 5.10) it is likely that larger-scale farmers in Archaic and Classical times did not regularly exploit sloping land by means of agricultural terracing. This is not to say that terracing was *never* used in classical antiquity, but rather to emphasize that there is little positive evidence to support the widespread assumption that terracing was a technique routinely employed by farmers in the Archaic and Classical periods.

There are serious taphonomic and methodological issues which confront attempts to date terracing in the archaeological record. Price and Nixon's (2005: 670) list of criteria for dating terrace walls 'in roughly descending order of strength' consists of:

1. Datable material in fill.
2. Age of trees on terraces.
3. Construction style of terraces.
4. Same construction style as adjacent ancient structures.
5. Terraces built against ancient structures.
6. Extent and type of lichenization of terraces in relation to the extent of lichenization of adjacent structures.
7. Extent of degradation of a terrace.
8. System of terraces of area with ancient sites and no later constructions: 'relict landscapes'.
9. Antiquity likely on other grounds to be the (or a) period of greatest pressure on agricultural resources.

Even in cases where several of these criteria can apparently be applied, there is usually ample room for uncertainty. Though these techniques may help to pinpoint specific terraces as older than a recent, modern period, none, with the possible exception of number 1, is likely to provide a fairly precise date within a reasonably narrow chronological band (e.g. 'Archaic' or 'Classical'). Criteria 2 (age of trees) and 6 (lichenization) will undoubtedly depend on local climatic and environmental conditions, and unless, unusually, dendrochronological dating is possible, these techniques can only suggest that terraces might be 'old', without allowing the assignment of a specific date. Similarly, criterion 7 (extent of degradation) is likely to be dependent on local erosional, geomorphological and environmental factors, and cannot possibly provide a sound basis for precise dating to a specific period. Criteria 3 (construction style) and 4 (construction style being the same as nearby ancient structures) are notoriously subjective. Frequently, the masonry style of rural structures, particularly dry stonewalling, is constrained to some extent by the ways in which local stone fractures, so may look quite similar over a long time period. Often, too, the masonry 'style' of rural structures in general is quite rudimentary, and it would be unwise to push the significance of alleged stylistic 'characteristics' too hard without independent dating evidence for so doing. If terrace walls are genuinely very old they may well have been repaired and rebuilt over their lifetime. Sometimes such modifications can be distinguished, but not always, and in any case it is likely to be impossible to date them with any precision. The possibility that terraces may have been rebuilt or landscapes re-terraced repeatedly over a long period complicates our attempts to read their constructional history. Criterion 5 (terrace built against ancient structure) provides only a *terminus post quem* for the terrace, i.e. it could in theory date to *any* time after the ancient structure. Criteria 8 ('relict landscapes') and 9 (period of greatest pressure on resources) can hardly provide *evidence* for dating as they are founded upon the assumptions which most regional archaeological projects are trying to test in the first place, and leave open the possibility of circular argument. Landscapes can be utilized and exploited for agriculture in ways that leave no clear archaeological fingerprints, and there are well-documented examples to show that the apparent absence of cultural material from a particular period does not mean that the area was deserted or unused.

Even the excavation of terraces, which has only rarely been attempted, need not necessarily provide unequivocal dating evidence because of the problem of relating the stratigraphy of soil horizons to the wall structure itself. On Delos, Brunet and Poupet (1997; Brunet 1999) carried out small test excavations behind terraces which seemed likely to be ancient at three different points on the island; two in granite areas (Champs 1 and 2), and one (Champ 3) where the underlying geology was predominantly gneiss (Fig. 3.1). To date final publication of this work is not available (notably none of the soil analyses is included nor any soil micromorphology done), and the preliminary publications do not include section drawings or stratigraphical matrices, so it is hard to be certain about the precise details of stratigraphical relationships. Brunet and Poupet (1997: 778–9) report the find of what appear to be strata of disturbed soil which they interpret as ancient agricultural soils. In Champ 1, the deepest of these soil horizons (0.60 m in thickness), situated just above the underlying granite (itself at 1 m depth), was dated on the basis of sherds near the bottom of it to the sixth–fifth centuries BCE. It also included charcoal and archaeobotanical material, but no details of these finds are published. In the upper levels of this excavation, at 0.35–0.40 m depth, a soil horizon identified as ancient and including obsidian flakes is dated on the basis of the identifiable sherd material

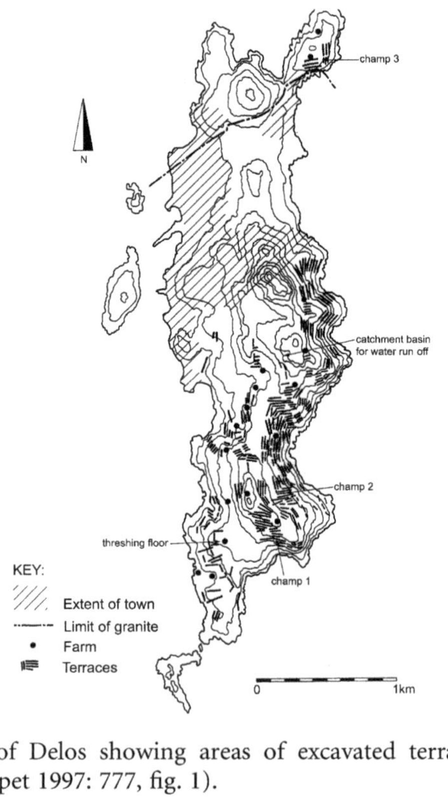

Fig. 3.1. Map of Delos showing areas of excavated terrace walls (after Brunet and Poupet 1997: 777, fig. 1).

to the Hellenistic period. At Champs 2 and 3, although numerous Hellenistic sherds were found in similar disturbed soil strata identified as ancient agricultural soils, the stratigraphy appears to have been less clear. Brunet (1999: 9) suggests that these Hellenistic soil horizons indicate a countryside abandoned around the end of the first century BCE.

Soil strata are hard to date because most of the sherd material incorporated in them is too worn to provide good chronological evidence. Moreover, one can never be certain that the datable material represents the *only* relevant period(s); at best it can only serve as a *terminus post quem* for the soil horizon since in theory the disturbed soil could have been in use at any point after the latest sherd material was deposited, incorporating pre-existing material from earlier periods. The other difficulty in this case is that there is no obvious stratigraphic relationship between these soil horizons and the walls. Did the soil strata discovered extend beyond or even underneath the terrace walls? What would excavation of the lower face (the 'front') of the wall have revealed? How can we be certain when the walls were built? They could in theory be either earlier in origin than the oldest of the soil horizons (dated at sixth–fifth century BCE) or more recent than the soil horizons dated to the Hellenistic period—there is nothing obvious that links the walls stratigraphically to any of the soil strata discovered. Fairly extensive areas of terracing are clearly documented by early twentieth-century photographs, where the lack of scrub and vegetation on them suggests that they were still in use at that time for cultivation (Cayeux 1911: 198, Fig. 100; 204, Fig. 107). Certainly, travellers in the nineteenth century described Delos as depopulated, but nonetheless reasonably fertile, cultivated in cereals by farmers from Mykonos and exploited by shepherds (Tozer 1890: 11; Bent 1885: 230). It is not beyond the bounds of possibility that the terraces excavated by Brunet and Poupet were exploited relatively recently. Clearly the archaeological investigations of these terraces are of the greatest importance, and their final publication should considerably illuminate our understanding of the terrace systems on Delos, and perhaps their dating, but as yet many questions remain unanswered. It may be that archaeologists need to develop and refine investigative techniques if we are to be able to date terrace walls with any confidence.

Most of the ancient literary and epigraphical references to field walls (now conveniently collected in Price and Nixon 2005: 686–91) are equally ambiguous. Price and Nixon (2005: 666) admit that the historical evidence for the use of terraces in antiquity is thin. Some sources might refer to terrace walls, but in virtually no cases can we be certain of their function. *Teichion* and *haimasia* (the specific word often thought to mean agricultural terrace wall) are both general Greek words for 'wall' and were used to designate a range of different kinds of walls. In most of the references collected by Price and Nixon (2005: 666) it is clear that *haimasia* refers to free-standing walls. Price and Nixon (2005: 690–1) have identified seven examples, which they believe use *haimasia* to indicate terrace walls, out of their corpus of thirty six references to ancient field walls of various kinds. The first of these (Text 27: Price and Nixon 2005: 690) is important since it appears in the *Digest* (10.1.13) in a passage in which the Roman lawyer Gaius is credited with quoting a Greek law about boundaries attributed to Solon:

Gaius libro quarto ad legem duodecim tabularum. Sciendum est in actione finium regundorum illud observandum esse, quod ad exemplum quodammodo eius legis scriptum est, quam Athenis Solonem dicitur tulisse: nam illic ita est: *ἐάν τις αἱμασιὰν παρ᾽ ἀλλοτρίῳ χωρίῳ <οἰκοδομῇ ἢ> ὀφρύγην, τὸν ὅρον μὴ παραβαίνειν· ἐὰν τείχιον, πόδα ἀπολείπειν· ἐὰν δε οἴκημα, δύο πόδας· ἐὰν δὲ τάφρον ἢ βόθυνον ὀρύττῃ ὅσον τὸ βάθος ᾖ, τοσοῦτον ἀπολείπειν· ἐὰν δὲ φρέαρ, ὀργυιάν. ἐλαίαν δὲ καὶ συκῆν ἐννέα πόδας ἀπὸ τοῦ ἀλλοτρίου φυτεύειν, τὰ δὲ ἄλλα δένδρα πέντε πόδας.*

Gaius in the fourth Book on the Law of the Twelve Tables. We must remember that in the action for regulating boundaries we should observe the rule that was formulated roughly on the model of the law Solon is said to have passed at Athens; there it is stated: 'If someone builds a *haimasia* or an embankment (*ophrygê*) next to someone else's land, he should not cross the boundary; if he builds a *teichion*, he should leave a gap of one foot, if a building two feet; if he digs a trench or a pit, he should leave a gap equal to the depth; if a well a gap of one *orguia*; he should plant an olive tree or a fig tree nine feet away from the other person's land, other trees five feet away.'

It is difficult to attribute this passage to Solonian law with any certainty (Blok and Lardinois 2006) set as it is in a compendium of

Roman law compiled in late antiquity. The lack of any original context for the passage is compounded by a problem with the text at the critical spot, obscuring the meaning of *haimasia* and certainly not illuminating the meaning of *ophrugê*. This is a rare word which occurs otherwise only in the late antique Lexicon of Hesychius. Its rather broad definition, 'χῶμα, λόφος, αἱμασιά' ('mound, crest, retaining(?) wall') does not inspire confidence in Price and Nixon's (2005: 668–9) claim that the Digest text distinguishes between *haimasia* or embankment (*ophrugê*) and *teichion*, with the intention of differentiating the appropriate relationship to a neighbouring property of a terrace wall, an embankment without walling, and a free-standing wall. This level of technical precision in the use of the three terms is difficult to prove with certainty in a fragmentary and obscure text which could be interpreted in other ways. *Ophrugê*, as Price and Nixon (2005: 669) point out, is related to the word *ophrus* (eyebrow, brow), but the latter word was regularly used metaphorically to refer to landscape features which stick out or overhang. Embankment is one possible interpretation, but not the only one. It could, for example, refer to an overhanging wall, a boundary wall on a ridge, or some other kind of feature which we do not understand. Similarly, *haimasia* might be an agricultural terrace wall, but it is not possible to be certain. It is also possible to read the passage as two separate but related legal statements, the first making the point that constructions of walls and features that stick out in general should not transgress a neighbour's boundary, followed by specific recommendations on the amount of space to be left between particular features and the boundary: walls, buildings, trenches or ditches,[5] holes, wells, and crop trees. However, given the state of the passage it is impossible to depend too heavily on the interpretation of specific terms in it.

The other passages cited by Price and Nixon (2005: 690–1) as evidence of the word *haimasia* used to mean 'terrace wall' are all third century BCE or later. All but one are epigraphical, and appear in

[5] Price and Nixon (2005: 690) translate *taphros* as 'grave'. However, the basic meaning of the word is 'trench' or 'ditch', and if we are correct in interpreting the context to which this passage refers as largely rural and agricultural, then 'trench' is probably more accurate—the word in fact never appears to mean 'grave' in other contexts (sv. *taphros*, *LSJ*), and Price and Nixon may have confused it with *taphos*.

inscriptions from Attica (text 28—*IG* II[2] 1322 [Rhamnous] and text 29—Agora 19.I4b [near Sounion]), Anaphe (text 30—*IG* XII.3.248), and Mylasa (text 31—*SEG* 2.545, *Inschr. v. Mylasa* 255 and text 32—*Inschr. v. Mylasa* 253). These passages could be referring to agricultural terraces, but other possible interpretations cannot be excluded where, as is usual with inscriptions, we cannot understand the full context. In some ways the most convincing is text 33, a brief entry in Hesychius's *Lexicon* of the fifth-sixth century AD under ὄφρυα: 'the lofty and up-lying fields. Some people for *haimasiai*'.

In summary, I am not arguing that there were no agricultural terraces in Archaic and Classical Greece or that Greek farmers never used terracing. Undoubtedly, some of the land that even rich farmers would have cultivated, especially that used for tree crops, would not even have existed as fields unless the rocks had been removed and terraces had been built. Terracing may have been a technique practised more often by poor than by wealthy landowners (see Chapter 5.10 below). The evidence for Archaic and Classical terracing is weak because 1) it is almost impossible at the present state of our knowledge to date archaeological examples of terracing with any precision or certainty; and 2) because there are virtually no secure and unambiguous references to agricultural terraces in written sources of any ancient period. It is possible, of course, that terracing was so much taken for granted in classical antiquity that it is never mentioned in written sources, but this is a dangerous argument from silence. Too often, scholars have tried to date terraced landscapes to their preferred period by wishful thinking, with little sound evidence to support their arguments.

Free-standing walls are documented much more regularly in both written and archaeological sources (Price and Nixon 2005: 666–8). These appear to have served a range of functions, including keeping livestock in and intruders out. One interesting example appears in a lease from Amorgos (Rhodes and Osborne 2003: 282–6, no. 9; *SIG*[3] 963.10–12; cf. Foxhall 1996: 48–50) where stipulations are set for the building and maintenance of free-standing walls with prickly bushes set on top for added protection (Rhodes and Osborne 2003: 283 blur the impact of the verb φράξει in their translation). Sometimes they served as boundaries, but not all boundaries were walls.

3.4.3. Drainage and Irrigation

Wells, cisterns and water-lifting devices also improved the cultivation potential and the value of land by facilitating the planting of crop trees and the watering of livestock, as well as permitting small-scale irrigation. Many such installations are found in the course of field survey: they have occasionally been known to be particularly hazardous discoveries for unwary field walkers. The Attic Stelae record the sale of three shadufs (water-lifting devices) and a large stone trough to receive the water (*IG* I^3 422.187–90). Wells or cisterns from which the water is lifted are not mentioned, but must be presupposed. Unfortunately the price is not preserved, and interpretation of the installations here is not straightforward. There are two likely possibilities, both of which take account of the fact that the water source is not mentioned: 1) that the shadufs have been dismantled and the buyer was going to reassemble them elsewhere; or 2) that these installations were sold off separately from the field in which they were located (they are apparently designated as 'in the field at Xypete', rather than being listed along with the sale of the field itself). Of course, the results of the biggest garage sale Athens had ever seen are not known, but it is perfectly possible that the land and the shadufs came to have different owners, if indeed they had not had different owners in the first place.

Larger-scale operations than those of a single household are also occasionally documented. One of the most intriguing appears in an inscription (*IG* XII.9.191) from Eretria in Euboea dating to the fourth century BCE publishing the terms of a large and complex contract between a wealthy citizen and large-scale landowner, Charephanes, and the polis, with mention of the local landowners affected by the proposed works. The text is fragmentary and its translation and interpretation are not straightforward (Prieto 2005: 253–8). However, the contract concerns a plan for the large-scale drainage of an area of marsh to create useable agricultural land. Chairephanes undertakes to pay for the full cost of the operation himself, and will apparently be organizing and overseeing the work since the polis gives him a tax-exemption on any wood and stone he will need to import for the project. He is bound by the contract to complete the project within four years, as dated from the year after the current archon, so at the time of the inscription work has probably not yet begun. In return Chairephanes will be entitled to lease the land (presumably from the polis) for ten years at a rent of thirty talents. This suggests that the land was in the territory of the city but was not owned by individual landowners. It might instead have been used as a resource for grazing, hunting, and gathering by the community as a whole. The price of the rent, ten talents per year, suggests that this was a large piece of land—one might suspect that Chairephanes was getting it at a bargain price rent for the first ten years.

Lines 18–20 make it clear that Chairephanes will have to run channels through privately-owned agricultural land to complete the project, and will have to compensate landowners for this. He is also required to make sure that in the process of drainage privately-owned cultivated fields are not flooded, by channelling the water into a reservoir. This reservoir must be fenced, but text mentions building a gate and implies that local landowners will have access to this stored water when they need it (lines 25–7). Unfortunately, the text is incomplete at this point and the interpretation is therefore uncertain. The gate mentioned could be understood as an entrance through the fence that had been erected to allow access for farmers and their animals. Or, as Prieto (2005: 257) translates, it could be understood as a sluice gate which, when shut, allowed the water usually flowing into the reservoir to be directed onto agricultural land rather than filling the reservoir. Either way, it was clearly envisaged by Chairephanes and the polis of Eretria that this reservoir would provide a useful supply of water for livestock and for watering crops, in addition to the benefit of the extra cultivable land gained from draining the marsh.

An interesting feature of this inscription is the elaborate and detailed provision to ensure that Chairephanes and his heirs fulfil the terms of the contract. Moreover, the terms were to be agreed with all the citizens individually by the swearing of an oath, written out in the inscription, in the sanctuary of Apollo Daphnephoros. Those who become adults and enter the citizen body in subsequent years are also to swear the oath as long as the contract lasts. There is equally elaborate contingency planning in anticipation of risks such as war, lawsuits, political action, incomplete or faulty workmanship, non-payment of rent, or the death of Chairephanes himself. These extraordinary provisions underpin the magnitude of the project, and perhaps also its unusual character. They might also imply that it was somewhat controversial, and that not all citizens could be counted on to back it in the longer term.

We do not know, of course, whether this project ever got off the ground, let alone whether it was ever completed. In its inspiration, this project is reminiscent of the undated but possibly ancient collecting reservoirs for water discovered and partially excavated on Delos by Brunet and Poupet (1997: 779–82; Brunet 1999: 27–42). Even today, some of these features fill with water in the springtime (Brunet 1999: 31, Figs 18–19; 33, Fig. 21). However, Chairephanes's project in Eretria appears to have been planned on a much larger scale than the Delian features. It is suggestive of the scale of project to which a wealthy landowner might aspire. However, it is simultaneously suggestive of the limits of what such a wealthy land owner (and perhaps his family) might undertake—in this case it is clear that that the scale was sufficiently great and the nature of the project (probably on 'common' land) was such that the cooperation of both the administrative structures of the polis and the members of the citizen body were perceived to be essential for its success.

The importance of irrigation for agriculture in Classical Greece has been much debated in the secondary literature, with the discussion generally focusing on a handful of literary references (e.g. Hom. *Od.* 7.112–30; *Il.* 21.257–62; Theophr. *CP* 2.2.1–4; 383; *HP* 2.7.1; 7.5.1–2). Hanson (1999: 60–3) perceives irrigation as part of a new package of techniques which characterize the 'yeoman farmer' emerging in eighth century BCE Greece and developing into the citizen farmer of the classical period. Although he recognizes that hydraulic works were always relatively small in scale, he considers them to have been an important factor in agricultural productivity. Hordern and Purcell (2000: 244–7) take a similar view, regarding irrigation as generally underrated in importance by scholars of Classical Greece. Prieto (2005) is similarly optimistic, and suggests that the 'division lines' at Metaponto were related to some kind of irrigation system. Krasilnikoff (2002: 54–6, 58) has argued more cautiously that many of the literary references regularly cited as demonstrating the common use of irrigation in Classical Greece, notably Plato's *Laws* (6.8.761A–B), simply demonstrate the importance of water to Greeks but do not provide evidence of the scale or degree to which irrigation was actually used. He suggests that there was in fact considerable regional variation from one city to another (Krasilnikoff 2002: 58–9), and this appears to be borne out by the available archaeological and epigraphical evidence. Undoubtedly, further archaeological discoveries along the lines of those at Metaponto (Prieto 2005) will help us to understand better the extent to which Greek farmers did or did not use irrigation.

3.4.4. Labour and Land

The most important component in the long-term productive capacity of land was labour. For large-scale Greek farmers, this effectively meant the purchase and maintenance of slaves. In a semi-arid Mediterranean ecosystem, labour is more than simply a means of cultivating and utilizing land. It is only via large inputs of labour that usable land for cultivation can be created from Mediterranean forest or scrub in the first place, and labour was always a major limiting factor for the amount of cultivation farmers could undertake (Foxhall 2003; 1996; cf. Jongman 1988: 89). Once cultivable land is created and made productive, continuous inputs of labour are vital for its continued existence. For cultivation, even for tree crops, land must be kept cleared of scrub, and often large numbers of rocks must be removed as well. Once cleared, soil had to be kept in place and run-off controlled by the construction of walls, trenches, ditches, and so forth. This was even more important on slopes. These fixtures then had to be continuously maintained to prevent erosion, and the land itself needed constant attention to prevent the regeneration of scrub growth. Similarly, in marshy areas, usable land could only be created if labour was first invested in draining it, and then in maintaining drainage systems (see section 3.4.3; Osborne 1987: 41, 44, 202–3; Jameson 1977: 128; Horden and Purcell 2000: 244–7).

Large landowners with numbers of slaves thus held a considerable advantage over small farmers in that the former could literally create usable farmland, making a permanently productive resource out of that which would otherwise have been wild grazing. Small farmers would not usually have had the labour reserves available within the household to develop land to the degree of productivity that large-scale farmers could have done. In this light, the numerous *eschatiai* attested as being under cultivation in the later fifth and fourth centuries BCE are significant. If they are indeed land that was marginal in some sense (Lambert 1997; Lewis 1973: 210–12; Jameson 2002; N. Jones 2004: 26–7), it is interesting to see how frequently they turn up in the hands of the wealthy.

The so-called *Rationes Centesimarum* inscriptions offer a particularly interesting example from the second half of the fourth century BCE. These are fragments of probably four stelae erected on the Acropolis recording a one per cent tax (though it is not clear who paid), most likely given to Athena and the Other Gods, on the purchases of land sold by demes and other corporate bodies (Lambert 1997: 1–5).[6] It is possible that these bodies were encouraged to sell land over which they had at least nominal control to private buyers in order to raise revenue for the state (Lambert 1997: 283). Whatever the motivation for the sales, it is clear that the Athenian elite comprised a major component of the purchasers (Lambert 1997: 244). Much of the land appears to have been 'marginal' or at least undeveloped—many of the plots are described as *eschatiai* and marshland is also mentioned. Generally it seems to have been sold off in large parcels, to judge from the prices paid, to members of the bodies selling it. These do not seem to have been ordinary commercial transactions (Lambert 1997: 232); prices appear to have been set in units based on 12.5, and the suggestion that this represents a purchase price calculated from the annual rent multiplied by 12.5 is persuasive (Lambert 1997: 263). This could imply that in many cases these plots were being sold to sitting tenants, and it seems likely that often they were sold to associates of the officials in charge (Lambert 1997: 248–50). The generally undeveloped character of the land, and the fact that in at least one case a public resource (threshing floors) is sold, might well suggest that much of the land sold on was 'common land'; within the control of the deme and in

[6] It was suggested by Lewis (1973; cf. Osborne 1985b: 56–9) that these inscriptions disguise leases but do not represent genuine sales, but Lambert (1997: 257–63) makes a convincing case for the view that the land was actually sold.

theory previously available to all, but not intensively utilized by any particular private citizen (Lambert 1997: 228–9, 238–9).

Rich slave-owning proprietors would have been the ones most able to mobilize the labour to develop these kinds of plots effectively for the generation of income. Phainippos, as he is represented in Demosthenes 42 (de ste Croix 1966), may have been one such proprietor. As can also be seen on a much larger scale in Roman Italy, one of the obvious ways of developing some of this property was by preparing it for and planting tree crops, particularly olives and vines. Another way, even before it is developed for agriculture, is to exploit it for forest products: timber, brushwood, resin, charcoal, and lime. In the recent past in Greece, such forest products have been a major source of cash income (Forbes 1996). Adeimantos's 'oakery' and 'pinery' probably represent the exploitation of marginal land for forest products (*IG* I³ 430.1–4).

Scholarly debate on the role of slave labour in Greek agriculture (Jameson 1977, de Ste Croix 1981; Wood 1983; 1988; S. Morris and Papadopoulos 2005) has often focused on smaller-scale farms. However, Wood's (1983: 15, 31) attempt to cast doubt on the long-held view that slaves were the predominant form of labour on 'large estates', and that, 'large estates' themselves were relatively insignificant is probably misguided. The fundamental flaw is her misunderstanding of the terminology of slavery (Wood 1983: 3–4). Despite her arguments to the contrary, *oiketês* is well-documented as a word regularly used to mean 'slave' and in the context of classical Athens it rarely means anything except 'slave' (Jameson 1992; Pomeroy 1994: 65–7, 314–17. For a good review of the history of changing views on Athenian slavery, see Roberts 1994: 262–76). As Jameson (1977: 137) long ago pointed out, the term refers, 'not to the slave's role as a domestic servant but to his place in the *oikos*, the household'. Slaves are not specified in the literary sources as hewers of wood, drawers of water and diggers of trees for large landholders because there was no need to specify it: everyone knew that slaves did the work for rich men, and this was the background assumption of Xenophon's *Oikonomikos*.

Many of Wood's assumptions concerning the significance (or not) of large estates, also reflected in later work (Miekle 2002: 242, quoted above), are problematic. She argues (Wood 1983: 29–31) that tenancy and sharecropping were the predominant modes of exploitation for large landholders. In fact, almost no evidence for private tenancy exists in classical Athens (Osborne 1988). The one well-documented case, Lysias 7, discussed in detail in Chapter 5.9 is almost certainly peculiar. Here the land in question may have been rented out because it was being 'held in trust' for the exiled oligarch Periander (the owner before it was confiscated by the state) by his friends, should he or his descendants ever return to Attica (S. Humphreys, pers. comm.). Wood (1983: 31) distinguishes between the scattered holdings of large landowners and large consolidated estates which might have been worked by slaves, for which there is in fact no evidence. In light of all the evidence for the fragmented holdings of wealthy households it seems more likely that scattered plots were worked by slaves based (for at least most of the year) at a country house, which might even be in a village. Most recently, S. Morris and Papadoupoulos (2005) have argued on the basis of Demosthenes 47 in combination with visible archaeological remains that farmhouse towers were used as secure accommodation for slave workers on isolated areas of agricultural land owned for the most part by wealthy land owners. The independent arguments of Osborne (1992a) and Foxhall (1992) suggest that under ten per cent of the Athenian citizen population owned or otherwise controlled at least one third, and possibly nearly half, of the cultivated land of Attica. That wealthy householders and their workforces were insignificant in the agrarian economy of ancient Attica and Athens must certainly be incorrect.

3.5. TREES AND TIMESCALES

One of the most important improvements regularly made on land to increase the value was the planting of trees, including vines[7]: 'But this man's patrimonial land, gentlemen, my father planted in trees and cultivated and made it double in value' (Is. 9.28). References to the

[7] Greeks classified vines as trees: see Theophrastos, *CP* 3, passim, Xenophon, *Oec.* 19.12, with the description of setting vine cuttings described first under the general category of 'tree planting'.

planting of trees are ubiquitous in the written sources. Theophrastos's botanical books are more concerned with aspects of tree planting than any other aspect of agriculture, and Book 3 of the *Causes of Plants* is largely concerned with the effects of different soils, climates, moisture conditions, and cultivation techniques on young trees. It is also a major part of Ischomachos's 'seat of the pants' farming in Xenophon's *Oikonomikos* (19.1–14), although he never actually says anything about the cultivation and maintenance of established trees. Clearly, one of the ways Ischomachos and his father are supposed to have improved the land they purchased was by planting it in trees, having bought it ἀργὸς καὶ ἀφεύτος, 'uncultivated and not planted in trees' (20.22). By the time they had finished with it, so Ischomachos claims, 'we made the land increase in value many times over the original price' (20.24). It might be asked why, in fourth century Athens, when population remained relatively high (even if lower that in the late fifth century) and good agricultural land was apparently at a premium, was there any land which remained undeveloped? Xenophon, via Ischomachos, provides a partial, if rather moralistic, answer (*Oikonomikos* 20.22): 'either through neglect (*ameleian*) or through incapacity (*adunamian*)'. Although the moral aspect of this comment was important for Xenophon's philosophical argument in the treatise (Finley 1999:19; Pomeroy 1994: 254–5, 322–3; Osborne 1987:17–8), there is a more practical element in the background as well: planting trees demands a surplus of available labour. 'Neglect' is effectively the same as a low labour input (e.g. Dem. 55.11); 'incapacity' is also the inability to provide labour. Hence we are back to labour as key element: for tree planting it was probably the major, though not the only, expense.

The passage from Isaios quoted at the beginning of this section also highlights another important feature of tree planting as a land improvement strategy: it operated in the long-term. How long depends on the kinds of trees planted. The olive is notorious for the length of time it takes to come into full production, perhaps twenty five to thirty years, although a small return may be made after eight to ten years, with luck. Once established, however, the trees live for a very long time. If a landowner planted olives on land that he intended to keep, it was his children and his grandchildren who would harvest the fruit.

Vines, on the other hand, give respectable returns after only four or five years, with the plants coming into full production after ten to fifteen years. This is no doubt at least part of the reason for their evident popularity as an 'investment' crop. The literary sources contain more discussions of the planting and cultivation of vines than of any other perennial cultigen. Although they are relatively short-lived, they are easily replaced. Again, the numerous and detailed discussions of vine planting and grafting in Theophrastos and, less completely, in other ancient Greek writers (even the few preserved fragments of Androtion make it obvious that vine propagation was a major concern of his work) show that large-scale landowners had propagation techniques down to a fine art. Such discussions also highlight the high labour inputs required for the systems of vine cultivation used on the estates of the rich (Hanson 1992). Other fruit trees (most importantly, figs, almonds, apples, and pears) come somewhere in between these two extremes of productive timescales and labour costs.

The technical detail of the planting and cultivation of trees, especially the olive, will be covered in more detail in Chapter 5. However, it was clearly one of the most important means of increasing both the long-term value and the productivity of land.

3.6. OLIVES AND OTHER CROP TREES ON LARGE-SCALE LANDHOLDINGS

To summarize the implication of the preceding sections, the crop trees owned by wealthy households were only one element of a complex and diverse range of activities, including non-agricultural ones in some cases. Given the variety of productive activities within households combined with the fragmentation of landholdings, it is not surprising that permutations of 'mixed farming' were the norm. The main aims of such households in the acquisition, attempted acquisition, and planting of crop trees (with the possible exception of vines) were more likely to be 1) both short-term and long-term *domestic* consumption needs; and 2) the investment of surplus labour resources to increase the value of land, either for the future use of children, or for more immediate purposes. Specialized production for sale was not necessarily a major consideration, though sale was one way of disposing of surpluses, when conditions were favourable.

In light of the mythology of the olive and its symbolic importance in Athens, it is striking how little the olive features in the Athenian sources on agricultural practices. Solon's alleged restriction of Attic agricultural exports to olive oil (Plut. *Sol.* 24; see section 1.5) need not, of course, indicate regular surpluses of olive oil, though it has often been interpreted that way. In fact, during the period of the fifth through to the third centuries BCE, from which most of the sources date, there is little solid evidence for the large-scale, specialized production of Attic olive oil or trade in it.

It is well known that olive crops are highly variable from harvest to harvest (Pansiot and Rebour 1961:10–11, 211–12; Osborne 1987:45–6), with returns ranging from very good to very poor (see Table 3.1). Moreover, in the absence of irrigation, trees generally only produce a crop every other year at best. Trees tend to be synchronized over a wide area, having the same 'on' and 'off' years.[8] The sheer unpredictability of olives provides a major reason against specialization in them. It is expectable that surpluses of olive oil and olives would have been sold and even exported from Attica when they occurred, and oil from the sacred olives (*moriai*) was certainly given away as prizes every four years, but the occurrence of surpluses was probably neither regular nor predictable.

In classical antiquity the unpredictability of the olive is exacerbated by the fact that the main source of labour for wealthy landowners was slaves. The labour requirements of established olive trees are erratic and seasonal. Pruning, for example, is not necessary or desirable every year, and increased labour input ceases to yield proportionally increased returns very quickly. At certain times, particularly the harvest, large amounts of labour are needed, while much of the rest of the time the trees need little or no attention. Moreover, the olive harvest conflicted directly with the main sowing time for cereals, which might mean that hired labour was difficult to obtain.

[8] This seems to be the case at least in southern Greece and southern Italy, though other areas of the Mediterranean region may be different (Forbes and Foxhall 1978; and see Chapter 1).

Table 3.1. Projected returns per hectare (Kcal) from olives [oil], wheat, polycropped wheat and olives [oil], figs.

Olives

1 ha land, trees planted 10 m apart = 100 trees per ha (maximum)
Probable maximum yield oil: 3.4 kg per tree
= 340 kg/ha oil per biennium
= 3,060,000 Kcal/ha (Pellett and Shadarevian 1970—900 Kcal per 100 g food energy value)

Wheat

(Theophrastos CP 3.11.1 ff; Forbes 1982: 334–50)
1 ha land planted in wheat, alternate year fallow assumed (probably unjustified) extrapolated sowing rate of 150–200 kg/ha (between Columella and modern Methana).
estimated yield of 5:1
= 750–1000 kg/ha wheat per biennium
= 2,655,000–3,540,000 Kcal/ha (Pellett and Shadarevian 1970—354 Kcal per 100 g food energy value)

Polycropped olives/wheat

1 ha land with 60 olive trees (probably low), leaving ca 0.75 ha available for wheat, biennial fallow assumed (probably justified). Extrapolated sowing rate of 150–200 kg/ha, as above. Estimated yield 4:1 for wheat; 3.0 kg oil per tree for olives, allowing for reduced returns because of competition for water and soil nutrients.

= 180 kg/ha oil	}	= 1,620,000 Kcal
	}per biennium	
= 450–600 kg/ha wheat	}	= 1,593,000–2,124,000 Kcal

= 3,213000–3,744,000 Kcal per biennium (Pellett and Shadarevian 1970—as above)

Figs

1 ha land with 100 fig trees (maximum)
50–(maximum) 100 kg dried figs per tree per year
=5000 kg/ha figs, annually (= 10,000 kg/ha per biennium)
=15,150,000 Kcal/ha per year(Pellett and Shadarevian 1970—303 Kcal per 100 g food energy value)
=30,300,000 Kcal/ha per biennium

In a regime where slaves provide the primary labour force, practical constraints suggest that olives are best combined with other crops.

On the other hand, the relatively high domestic consumption needs of wealthy families for olive oil and its high prestige value both for food and non-food uses (see Chapter 4) provided an impetus for large landowners to grow olives. Precisely because returns were unpredictable, quite substantial numbers of trees would be needed to ensure *domestic* supplies. Disposal of surpluses is unlikely to have been a problem. Because of its great prestige value, it is likely that demand for olive oil (especially high-quality oil) was highly elastic: if supplies generally increased because of a bumper harvest, demand might rise nearly as fast as price fell. Since it keeps adequately for six or more years, surpluses could be held in storage by those who could afford to invest in the storage facilities. Hence, the occasional very large crop would not necessarily present difficulties for large-scale growers.

Figs, though similar in their cultivation requirements to olives, give much more regular returns. Dried figs were a major staple, especially of slaves, poor people, or for everyone in wartime (Foxhall 1993a: 41). Productivity of the trees is very high: a large fig tree could be expected to produce 100 kg or more of dried figs annually. For smaller, less-productive trees, around fifty kg of dried figs per year could be expected.[9] With a food energy value of 303 Kcal per 100 g (Pellett and Shadarevian 1970), this means that land planted only in fig trees (with no other crops grown) would produce around 15 million Kcal per ha annually, compared with, for example, olives, producing around 3 million Kcal per ha, or wheat, producing around 2.6–3.5 million Kcal per ha, biennially (see Table 3.1). Minimal processing, in terms of both time and equipment, was needed. Trees did not take so long to come into full production as olives. Moreover, the fig harvest came at an otherwise slack time of year in late summer. All in all, they seem to have been prominent in the Attic countryside and hitherto underestimated in importance.

For rich householders engaging in agricultural speculation (to the extent that any may have done so), vines must have been tempting. It is suggestive that the planting and cultivation of vines is the major concern of most of the sources we possess and even some of those we do not (e.g. Androtion). For a start, vines have cachet: they are a crop with a high cultural value producing (or at least so the owner hopes) a high-status product. They are also highly compatible with the use of a slave labour force: work can be spread reasonably evenly throughout the year, with only one major peak at the vintage. Compared with

[9] These figures are calculated using modern data from Methana and Kranidhi, Peloponnese, Greece (H.A. Forbes, unpublished field notes). For olive oil yields, the calculations are based on figures that would probably represent maximum possible production per ha for oil. For figs, the calculations for large trees are near maxima.

other fruit trees, they come into full production quickly, within a couple of years.

The Thasian farm of Adeimantos, inventoried on the Attic Stelae (*IG* I^3 426.44–56; see Table 2.1), provides a 'real life' example of a specialized vineyard, though not, interestingly, in Attica, but on an island famous in antiquity for its wine (Brun 2004a: 94). The price is unfortunately not preserved, but from the items which are recorded on the stone, the farm seems to have been quite large. Since this section of the stone is quite damaged, it is probable that other items were recorded as well. Besides the house and the land, there is a named slave (presumably the bailiff-manager), large numbers of good and bad *pithoi* with lids, and 590 amphorae of wine with a capacity of 3 *choai* each. This is the equivalent of nearly 6000 (5098 to be exact) litres of wine in store. Such a large amount must surely have been produced with sale in mind, especially given the reputation of Thasos in antiquity for wine production, a reputation which is borne out by the many finds all over the Mediterranean of stamped amphorae from the island. The farm that Adeimantos owned ought to have produced a handsome profit.

Wine producing, however, is a risky business. Under pre-industrial conditions quality control is difficult. Vines are highly sensitive to inter-annual variability in weather conditions. This affects not only the quantities of grapes produced, but also the sugar content and thus the quality and flavour of the wine produced (as well as the quantity). Nothing will completely eliminate the effects of weather on the quality and quantity of wine produced from a vineyard (as modern chateau owners know to their cost), but high inputs of labour and planting locations that cover a variety of micro-environments can sometimes mitigate the worst effects of short-term meteorological variability (Theophr. *CP* 3.11.1 ff; Forbes 1982: 334–50; Hanson 1992).

Large numbers of vines, then, were not practicable for a small-scale, subsistence-based farming operation. But for large landowners, especially when the main form of agricultural labour was slavery, vines could have been very profitable. If labour was going to be invested and trees planted to improve land, vines would have been the most obvious crop choice for high returns in the short- or medium-term. On large landholdings vines and even figs may often have taken precedence for proprietors over olives. But, given the general trend toward mixed farming, the unpredictability of olive production and the high domestic needs for olive oil of wealthy proprietors, it is likely that most large farmers had at least some olive trees, possibly quite a few in most cases.

3.7. ARBORICULTURE AND ITS BY-PRODUCTS: ANIMAL HUSBANDRY

Arboriculture within mixed farming regimes offers an excellent fit with small-scale animal husbandry. The useful products of animals: meat, manure, milk and dairy products, skins, wool and hair, horn and bone, and traction, are well-known and well-documented in ancient Greece. Sheep, goats, donkeys, pigs, and cattle can be fed on many of the by-products, including the leaves of pruned branches of all fruit trees (the olive among them, Theophr. *HP* 1.12.4; 4.8.13; 4.10.7; 9.16.1), the press cake from wine and oil processing, and fruits spoiled or damaged by pests and diseases (Foxhall 1998c). This last method also has the advantage of keeping fungal infestations and insects in check by removing the sources of infection from the vicinity of the trees and destroying them, thus helping to prevent re-infection via eggs and dormant spores in the next year. In addition, grass and weeds growing under trees can be kept in check by animals (especially sheep and goats), for whom such areas provide a welcome source of grazing. Once the animals have eaten the leaves, branches from pruned trees and vines also serve as fuel for domestic use. Certainly in the recent past in Messenia vine prunings were considered to be a particularly desirable fuel for pottery kilns since they burn quickly but do not leave much residue, while olive press cake was also thought to be an excellent fuel for kilns (Matson 1972: 219; cf. on prunings Theophr. *HP* 5.9.6). Press cake can be stored for use in times when other fodder and grazing is in short supply, a practice documented as early as the Neolithic period (Galili *et al.* 1997, see Chapter 1.3) as well as in the Roman agricultural writers (Cato *RR* 11.1; Columella 6.3.4, 8) for wine press cake. Alternatively, olive press cake can be used for domestic fuel. Animals, then, utilize by-products of olive cultivation in particular and arboriculture in general, which humans cannot consume directly. The animals in turn produce useful products, assist in the battle against weeds, pests and diseases, and via their dung improve agricultural land.

3.8. MIXED FARMING: FLEXIBILITY AND OPPORTUNISM

The various mixed farming regimes common in the ancient Greek world enhanced the flexibility of farmers and fostered the mentality of opportunism that we have seen in the economic and business activities of wealthy, large-scale landowners. The symbolic and political importance of land stands alongside its economic importance as the primary means of production. Land also constituted a major element in the resources to be passed on to children, so long-term improvements enhanced that resource and thereby the stability of the total resource base of the household (and the future households generated by it) over time. Most of these improvements are relatively small in scale, constrained as they were by the social and economic framework of the household, and the financial and labour resources on which it could draw. Larger or more specialized improvements or 'capital investment' might also have been constraining in other ways. They would have tied up too great a proportion of available resources and limited flexibility—the capacity to change course in the face of changing circumstances within an environment, in the broad sense, that was perceived to be volatile and unstable. However, that does not mean that the impact of these small-scale improvements was insignificant, or that the aggregate impact of many households pursuing such strategies was small.

4

The Domestic Consumption of Olive Oil

4.1. THE CONSUMPTION OF OLIVE OIL BY WEALTHY HOUSEHOLDS

The domestic consumption needs of rich households in ancient Athens for olive oil falls into three basic categories: food, lighting, and personal cleansing and adornment. This last category includes the use of oil in the gymnasium as well as the use of oil for perfume (Raftopoulos 1995; 1996; Brun 2000). This chapter attempts to quantify these demands roughly, and the results are summarized in Table 4.1. Obviously, even a very crude estimate must depend heavily on the judicious use of comparative and 'proxy' data, and the resulting figures are inevitably very approximate. Nonetheless, the exercise is useful for the purpose of making a guess at the levels of production for which such households aimed, and thus for building the understanding of the economic, cultural, and nutritional significance of the olive in ancient Athens and Attica, and perhaps more widely in the Greek world, although there are bound to have been significant regional differences in consumption patterns.

A central problem for any attempt to quantify the use of oil in ancient times is that olive crops are notoriously unreliable and unpredictable from year to year. Hence, in order to secure domestic consumption, households might need substantial numbers of trees, which in some years would produce far more oil than a single household could consume. The result might then be surpluses which could be stored for times of shortage, sold for cash, or otherwise 'invested' in social and/or political relationships (Halstead and O'Shea 1989).

Table 4.1. Summary of estimated levels of domestic consumption of olive oil in wealthy Athenian households.

Oil for food	25–35 kg per person per year (modern: 50 kg person/year)	100–200 kg oil per household per year
Oil for bathing (including perfumed oil)	1.5 L person/year (women) 5–10 L person/year (men) (modern: 25 ml oil washes one six-year old girl)	10–20 kg oil per household per year
Oil for lighting	0.5 L oil per symposium 100 ml oil for 2 lamps for 2 evenings (modern: 1 lamp burns 7 hours, 50 ml oil)	90–110 kg oil per household per year
Total household oil consumption:		**200–330 kg oil per household per year**

This last element highlights an important aspect of olive oil consumption in Archaic and Classical Greece. As a highly processed product, with considerable cultural value attached to it, the consumption of olive oil was more than just expediency. It is probably no accident that the perceived quality scale for olive oil in antiquity was related to increasing elaboration of its production in terms of scarcity, difficulty of extraction, time and labour taken over pressing, and similar factors (see Chapter 6.1.2). Table olives, much easier and cheaper to make and needing no great amount of time, labour or equipment, never attained the highly-valued special status of olive oil in Greek culture.

The consumption of olive oil is a key element in the treatment of the body. For the Greeks, a strong and healthy (male) body in good condition was a moral reflection of the man himself, with important implications for an individual's political persona (Aesch. 1.31, 189). The literal embodiment (Hamilakis 1999) of specialness through the consumption of olive oil, internally and externally, often took place in social situations where that specialness was highlighted: formal and the symposium (oil for food, lighting and perfume), and the gymnasium (oil for cleansing and conditioning the athletic male body). In this sense the consumption of olive oil in all its forms does not merely enhance the social status of the consumer, it serves more basically as a key constituent of personal identity, through the body, of the consumer.

4.2. OLIVE OIL FOR FOOD

Although the diet of wealthy Greeks in the classical period certainly differed in many important respects from the rural Greek diet in modern times, many basic elements remain the same. The diet of almost everyone in classical antiquity, as in rural Greece up to the present, was largely cereal-based (Foxhall and Forbes 1982). It was almost certainly the case that beyond a certain level of wealth, the amount of grain that people ate decreased, to be replaced by more favoured foods, especially oil, dairy products, and meat (Foxhall and Forbes 1982: 70, 75). In the twentieth century, globally, the consumption of fats and oils rose with level of wealth more dramatically than the increase in intake levels of any other food, and this was a global trend (FAO/WHO 1973: 20–1).

On modern Methana, where a considerable number of olive trees are grown relative to other areas, householders estimate their consumption needs (food use and sometimes soap, not lighting) for olive oil at about 50 kg per person per year (Forbes 1982: 177). They actually store twice that amount if possible, to ensure subsistence supplies in the event of a shortfall or crop failure at the next olive harvest (Forbes 1982: 384–5; and subsequent fieldwork; Forbes and Foxhall 1995). This is almost certainly a generous estimate of consumption needs, which are over, rather than underestimated, to prevent shortages. Present day 'rules of thumb' may also date from a time when oil for non-food uses, especially lighting, were included in the estimate. However, it is clear from accounts of the 'bad old days' that in the past, people did not always have as much olive oil available for cooking as they would ideally have liked, or as they do at present. Similarly, Doorn (pers. comm.) working in Lokris, found that households estimated their annual consumption of olive oil at around 200–250 kg (assuming a family of 4 or 5, this also works out at around 50 kg per person per year).

Allbaugh (1953), working in post-war rural Crete,[1] closely documented the diet of his informants by assigning a researcher to

[1] In 1948, to be precise. This was a time of political turmoil and major food shortages in Greece, although rural areas were on the whole less badly affected than towns.

monitor a number of sample households' food consumption for a full week. These were then seasonally adjusted to allow for variations in consumption patterns throughout the year. In addition, he obtained households' own estimates of annual consumption of food. The results are interesting in light of the Methana data. Actual consumption of olive oil was about 31 kg (60 lb) per person per year (the monitored study and the seasonal adjustment results are the same). Householders' own estimates, however are much higher: around 45 kg. Allbaugh (1953: 107) gives his figures in pounds, and the figure of 100 lbs for households' own estimates may well represent a widespread informant response of '50 kilos' or '50 *okades*'[2] to the question of 'how much olive oil do you need for a year?'. Allbaugh (1953: 107) also noted that this estimate almost certainly includes oil for lighting and other uses (e.g. soap?). The estimate of consumption given by households themselves is thus very close to, if not indeed identical with, that of Methana and Lokrian farmers some 25 to 35 years later.

These figures must surely bear some resemblance to the consumption of wealthy Greeks in an olive producing region like Attica in classical antiquity. Less oil in total may have been available because production generally may have been lower (see Chapters 5 and 6), but consumption of oil by the rich in antiquity must have been at least as great as that of modern peasants.

Wealthy Athenian families in classical times would have entertained guests on a larger scale than peasant families do at present. Considerable amounts of olive oil may well have been used in the meals prepared for guests. This is clear from many extracts cited by Athenaios (1.6a, 1.7e, 2.64e, 2.65c, 3.110b, 3.117d, 3.126c–d, 4.169e, 4.170b—the details of these culinary uses are covered in more detail below). Moreover, the provision of oil (or scented oil) to guests, both as culinary condiment and for personal cleanliness and adornment, was part of the ritual of upper class hospitality. Athenaios (4.173d–e, and 1.5f–6a) quotes a comic reversal of this custom: the gourmet poet Philoxenos, who carried around his own seasonings and relishes. Scented oil was used not only as perfume, but also in cookery, as

[2] The *oka* was the Turkish and Venetian period weight standard, roughly equal to a kilogram.

medicine, applied internally and externally, and as a flavouring agent for wine (Theophr., *On Odours* 12, 32–6, 51–2). Whether scented and flavoured oils would regularly have been made within the household is not clear; certainly the manufacture of perfume was a specialist craft (Brun 2000; Harris 2002). An estimate of 25–35 kg oil per person per year (giving a household consumption figure of around 100–200 kg per year) is probably not excessive, and, were the oil available, households might have consumed much more.

However, there are further interesting and significant differences in ancient and modern attitudes toward the culinary use of olive oil, with implications for the role of olive oil in ancient diets. Most fundamentally, olive oil was plainly not considered an essential staple. Rather, with all its resonances of varying qualities, it was used as an *opson*, roughly translatable as 'seasoning, condiment, relish'—in other words, one of life's little luxuries to make food worth eating (Davidson 1998: 21). Athenaios (2.66f–67b) classes olive oil along with other commodities attributed with varying degrees of value and rarity, including fish pickle, pepper, pigs' brains, finches, blackbirds, and other delicacies. The consumption of such delicacies in excess, however, might also be represented as morally negative, as when Aischines (1.65) portrays Timarchos and his lowlife companions as hanging around the stalls in the market selling *opsa*—an indication of their fondness for lavish expenditure. Davidson (1998: 20–35) has argued that the term *opson* generally referred to fish, another food with profound and wide-ranging cultural meanings in Greek culture. Certainly fish and fish products were included within the category, and the word became more closely associated with fish in the Hellenistic and Roman periods. However, in Classical times and earlier, it is clear that the term included a wide range of foods, many of them in a 'semi-luxury' category, which were eaten as a flavourful relish to enliven the staple diet of bread and other forms of cereal. As such, *opsa* become emblematic of refinement, and thus the proper and moderate consumption of them defines the citizen man of means, while consumption in excess indicates moral degeneracy.

For elegant cookery, olive oil was essential. A fragment of Philemon, quoted by Athenaios (2.64e) vividly makes the point that when rich people ate boring, 'peasant' foods, they improved and upgraded them by using expensive, highly-prized ingredients:

Consider, if you wish, the bulb,[3]
How extravagantly it spends to keep up its reputation:
Cheese, honey,
Sesame, olive oil, onion, vinegar, silphium,
But by itself it's nasty and bitter.

Of this list of flavour improvers, only vinegar and onion are not relatively expensive items. Moreover, different qualities of culinary oil were distinguished. The best, called 'white oil', was believed to come from unripe olives and was felt to be both tastier and more digestible. Athenaios (2.66f.) describes Samian oil as the whitest of all.

The Hippocratic precepts on healthy eating are enlightening in this respect, and also suggest that attitudes to such 'semi-luxuries' were ambivalent. A considerable portion of *Regimen 2* (39–51) is devoted to the virtues and detriments of particular foods for promoting health and the 'good body' essential to the identity of the male citizen. The starting point and the longest section is a discussion of varieties of cereals and cereal products—obviously because they were the primary staple for most Greeks, rich or poor (Hipp. *Reg.2* 40–4). But olives are not explicitly mentioned at all, unless they were assumed to come under *opsa*, (Hipp. *Reg.2* 51) and olive oil is mentioned only with some disapproval, as heating, indigestible, and unhealthy in excess (Hipp. *Reg.2* 42, 51). Hippocratic views of the moderate diet which results in good health, as part of a sound and balanced 'lifestyle' (regimen, *diaitêmata*), falls to some extent within a moral and medical context. Within Hippocratic theory, foods were used to correct the individual body's imbalances, whether these were 'natural' to that individual or pathological. The ambivalent attitude displayed toward olive oil confirms its role as a non-staple, 'semi-luxury' food, healthy in moderation but not in excess. It may also suggest that wealthy Greeks ate more of it than their doctors thought was good for them. This is an interesting contrast to the positive view of olive oil when used externally for cleansing and bathing as part of a healthy regime (Hipp. *Reg.2* 58).

The consumption pattern of olive oil is quite different from that of table olives, ordinary varieties of which seem to have been a

[3] Various kinds of bulbs have traditionally been eaten in Greece, especially (but not exclusively) by the rural poor. Grape hyacinths (*Muscari spp.*) are a Lenten food even today.

widespread food staple, especially for the poor (Athenaios 2.56a–d; 2.60b–c). However, like oil, choice varieties of olives had a place as a relish or a '*meze*' (*opson* again) at the most elegant meals (Athenaios 2.68a: list of condiments/seasonings; 2.60a: selection of olives as a usual *opson*). In consequence, 'inferior' types of table olive, especially the black, wrinkled kind, preserved in salt, were believed to be nutritionally inferior and indeed bad for the digestion, compared with the more highly processed varieties gently crushed in a press (an expensive piece of equipment unaffordable to many poor farmers), then preserved in brine with herbs and spices.

Diphylos the Siphnian says that olives offer little nourishment and cause headaches; the black ones are bad for the stomach and heavy on the head, while the ones called 'swimmers' are easier on the stomach and are and astringent for the bowels; but black ones crushed are easier on the stomach (Athenaios 2.56a–b).

This provides a good example of how the nutritional value of a food was directly related to its status value from a Greek (and Roman) cultural perspective. Just as wheat and barley are nutritionally virtually identical, yet the preferred, more expensive cereal, wheat, was considered to be more nutritious (Foxhall and Forbes 1982; Hipp. *Reg.2* 42). Black olives, which were cheap and easily produced without the need for expensive ingredients or equipment, were considered nutritionally inferior to the green, highly processed varieties. In fact, it is probably the other way around (see Table 4.1). Per unit of weight, black olives are higher in fat, protein and calories, than green (Pellett and Shadarevian 1970: nos. 114, 115); in terms of other nutritional values they are identical.

4.3. OLIVE OIL FOR PERSONAL CLEANSING

The practice of rubbing down with olive oil (both scented and unscented) at the gymnasium after exercise is well-documented, and by the Hellenistic period the supply of oil for the town gymnasium was a regular benefaction of the wealthy. However, it is difficult to estimate the quantities used. Men and boys, for whom gymnasium exercise was a major part of their education, would probably have been the major household consumers here. But women, too, bathed with olive oil, though one would expect not as frequently as men. Sometimes perfumed oil seems to have been used literally as bath oil—that is, poured into the bath water to make it more pleasant and relaxing, for the sake of both health and comfort (Athen., 1.24c–d). Frequency of bathing is not a subject on which the ancient sources are very revealing, though Aristophanes' *Acharnians* 994–9 (quoted below, Chapter 7.3) suggests that the 'average Acharnian yokel' might be expected to bathe once a month! Upper class men and boys, on the other hand, seem to have gone to the gymnasium almost every day, and slaves were barred from participating in gymnastic activity (Aesch. 1.138). Indeed, the practice of rubbing down the body with olive oil was highly recommended as a healthy practice by the Hippocratic author of *Regimen 2* (55), especially in winter, since oil was considered to be 'warming'.

I found that it took about 25 ml of olive oil to rub down and clean one moderately grubby six year old child (height 115 cm; weight 20 kg). One might guess that for older boys and adults 25–50 ml per gymnasium visit might be a reasonably realistic estimate. Assuming an average attendance of 4 days per week through the year, this might consume five to ten litres of olive oil per person. If women consumed about 30 ml per week in bathing (and the amounts may have been higher), the annual rate of consumption would be about 1.5 litres per person. Household consumption of olive oil for bathing and gymnasium use must have varied enormously, but based on the figures worked out above, perhaps 10–20 kg per year might have been used by a wealthy family in this way.

4.4. OLIVE OIL FOR LIGHTING

Artificial light in the ancient world could also be defined as a commodity, the consumption of which was associated more with special occasions such as ritual (Parisinou 2000) than with utilitarian activities. Substantial amounts of artificial lighting at night would have been the prerogative of the wealthy in classical antiquity—in effect another form of conspicuous consumption. Though clay lamps were ubiquitous and obviously inexpensive, the oil would have been relatively costly and wick material had to be furnished as well. Small oil lamps do not give a great deal of light, and many would have been needed for evening entertaining in a wealthy household. A fragment of the fourth century BCE comic poet Antiphanes, preserved in Athenaios (4.170d–e), suggests that the slave who set up the tables for dinner guests also made ready the lamps. A small modern Greek metal lamp with a linen wick will burn for around seven hours at a fairly low level, using 50 ml of olive oil. I would guess that about ten lamps of this sort would be needed to give even dim illumination to an evening symposium, and more would be necessary to provide good light. This would consume around 0.5 litre of oil for one good party. Or, two such lamps would light a room dimly for perhaps two evenings (depending on the time of year). It is very difficult to arrive at an annual consumption rate of oil for lighting from these figures, since obviously households would have varied tremendously in their requirements. But for a wealthy family, which entertained frequently (maybe twice a month?) as well as illuminating other rooms in a large house regularly, perhaps 90–110 kg of oil would be an expected range of annual consumption. This also provides a reminder that the use of artificial lighting, especially in large amounts, was a highly visible form of conspicuous consumption.

4.5. THE DOMESTIC CONSUMPTION OF OLIVE OIL

In summary then, it is extremely difficult to estimate the amount of olive oil that would be 'necessary' for the total domestic consumption of a wealthy family in classical Greece. Partly, no doubt, it would have depended on how much oil they had in any particular year. But the potential for such a household to consume a very large amount of olive oil, were it available, is very great indeed. It is clear that olive oil was used on many socially significant occasions, and that its very presence was used to emphasize the special nature of an occasion, because of the high cultural value attached to olive oil, as well as the fact that it was an expensive commodity. Hence, there was a strong impetus for rich households to ensure that they had a reasonable number of productive trees, so that steady supplies of such a socially and nutritionally valuable product might be assured. The figures worked out above suggest that for a wealthy household in classical Greece (in an olive growing region, obviously), 200–330 kg of olive oil per year might be the range of consumption we might expect.

The complex social and cultural significance of olive oil, however, is even more important. Compared with table olives it was expensive to produce. On a spectrum of value (in both cultural and economic terms) from staple to luxury, it undoubtedly it fell towards the latter end, although it was a luxury that even quite poor people might afford in small quantities for special occasions (Foxhall 1998a, 2005). In contrast with table olives, it was not a staple food in any subsistence sense. Hence it takes on the cultural ambivalence which such luxuries and special treats so often do: at one level, it has the slightly wicked appeal of cream cakes in our world. The addition of scent and flavouring to oil is also important (often they were the same thing). The scents and flavours used, for example, rose, orris root, myrtle berries, cinnamon, and other exotic spices, further enhanced its status as a 'luxury' product. But scented oil was not only more luxurious, it also developed a further range of social, medicinal, and ritual uses. And both scented and unscented oil were used to treat both the inside and the outside of the body.

All of this suggests that we are seeing a cultural phenomenon which went beyond the simple association of 'luxury' and status markers. Olive oil is a product for which the range of value is closely entwined with the elaboration of processing techniques. Embodiment of olive oil became through its consumption became a critical constituent of personal and social identity, especially perhaps for men (feminine identities are much harder to see in the sources). In Attica, the existence of sacred olive trees (Chapter 5.9) dedicated to the goddess Athena, which were used for the manufacture of sacred oil given away as prizes in the Panathenaic games, adds a further dimension to the value of olive oil. This oil, literally a gift of the goddess in Athenian thought, must have built on a pre-existing set of symbolic values. Oil from sacred trees could be viewed as another kind of 'elaboration' in the processing, which both limited the quantity and increased the specialness. Yet, with this oil, the consumer also embodies a physical reification of 'being Athenian' in its most basic sense, and one which connects 'being Athenian' directly with the realm of the divine via a material food commodity.

The 'semi-luxury' value of olive oil may well have outstripped its economic significance in classical Attica. Levels of production of olive oil may have been relatively low compared to the high levels which have been generally assumed in assessments of Attic farming. Indeed, most of the olives grown may well have been made into table olives, which probably genuinely were an important staple. But this has perhaps been masked by the tremendous cultural significance of olive oil, perhaps most succinctly epitomized in the Panathenaic prizes and tiny aryballoi full of precious scented oil.

5

Cultivating the Olive

5.1. TREE PLANTING AND PROPAGATION: GENERAL ISSUES

The Greeks knew that most fruit trees do not grow true to type from seed (Theophr. *HP* 2.2.5), and even if they did, it would take an uneconomically long time to grow productive trees in this manner. Olive trees, like other crop trees, could be propagated in a number of different ways. The main methods used for almost all fruit trees were cuttings (Xen. *Oec.* 19.12), layers and grafting (usually bud grafting rather than twig grafting though both were known and used, see Fig. 5.1). For olives there was the additional possibility of propagation by cutting off and planting the ovules which develop at the base of the stems of mature olive trees. The olive, says Theophrastos (*HP* 2.1.4), can be propagated in more ways than any other kind of tree. All of these methods have their own advantages and disadvantages, and are better in some circumstances than in others[1]. Which was chosen depended upon:

1) The stock available to the farmer.
2) The particular place in which he intended to plant, in terms of:
   a) its location and accessibility;
   b) its nearness to water; and
   c) its soil type and exposure.

[1] In the following discussion I do not refer to Amouretti's (1986: 58–62) discussion of olive propagation, since I think she has seriously misinterpreted Greek terminology in light of Roman and modern French practices.

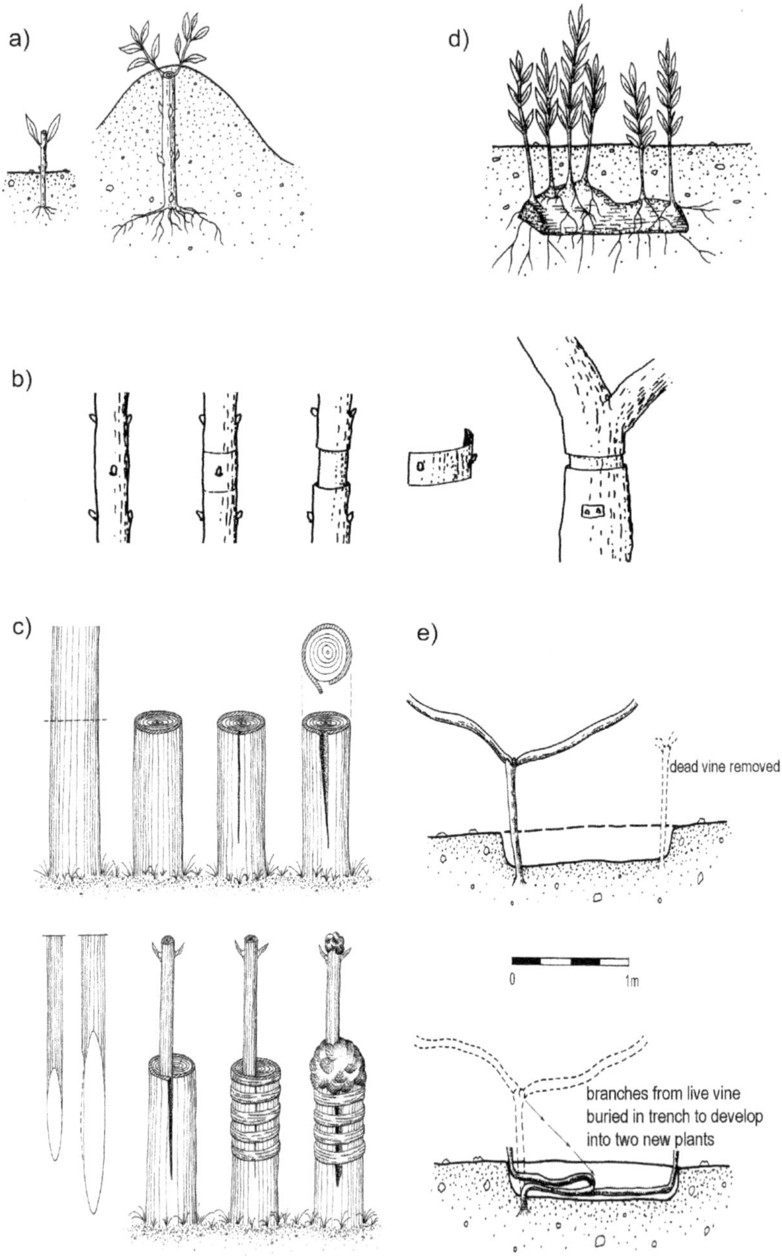

Fig. 5.1. Methods of fruit tree propagation: a) cuttings, b) bud grafting, c) top grafting, d) ovules (olives), e) layers (vines) (after Pansiot and Rebour 1961: 65, 69, 80, 73; Forbes 1982: 257).

3) The labour available, not only for the actual planting operation, but also for the maintenance of young plants once they were in place.
4) How many trees he intended to plant.
5) The speed at which he hoped the trees would come into production.

5.2. ROOTED AND UNROOTED CUTTINGS

Cuttings of various kinds were probably one of the most common methods of propagation, to judge from the number of references to the practice (Theophr. *HP* 2.2.4, 2.5.3 and *passim*; *CP* 3.5.1–5 and *passim*; Xen., *Oec.* 19.8–14). Cuttings could be planted out, either rooted (that is, with roots already established by cultivation in a nursery bed), or unrooted (that is, pieces of young branch wood simply stuck in to the ground (Theophr. *HP* 2.2.4.; *CP* 3.5.3–4; Xen. *Oec.* 19). A fair proportion of cuttings of either kind would not 'take', or would die before they became productive trees. The most regular cause would have been desiccation, but pests, diseases, sun scorch (Theophr. *CP* 5.9.1), and other accidents, such as voracious livestock or invading armies (Theophr. *CP* 5.17.6; Foxhall 1993a), also played their part in the high mortality rate of young trees. Although unrooted cuttings demanded less work and preparation to produce before the final planting out, since they were not tended in a nursery bed first, the success rate would have been lower and they would have needed even more watering, weeding, and cultivation to succeed than rooted cuttings, once planted out. Xenophon (*Oec.* 19.1–2) seems to describe in some detail the planting out of unrooted cuttings of vines (Xen. *Oec.* 19.12) in the course of the following dialogue between Sokrates and Ischomachos:

'I know that you have seen the kinds of trenches (*bothunoi*) they dig for trees (*phutoi*),' he [Ischomachos] said... 'Have you ever seen them deeper than three feet?'
'By Zeus not even five half-feet,' I said.
'Well, have you ever seen any more than three feet wide?'
'Not even two feet, by Zeus,' I said.

'Go on then,' he said, 'answer this one for me. Have you seen any less than a foot in depth?'

'Personally, not even less than three half-feet, by Zeus,' I said, for one would dig out (*exorutoito*) the cuttings (*phuta*) when they were being dug for cultivation (*skaptomena*), if they had been planted too close to the surface like that.'

'Surely then you already know this adequately, Sokrates,' he said, that they never excavate deeper than five half-feet nor shallower than three half-feet.'

'... Well then,' he said, 'can you recognize drier and wetter land on sight?'

'It seems to me at least,' I said, 'that the land around Lykbettos and land like it is dry, and that the land in the Phalerian marsh (*helei*) and land like it is wet.'

'Would you dig a deep trench (*bothron*) for a tree (*phutô*) in the dry land or in the wet?' he said.

'In the dry, by Zeus,' I said, 'since excavating a deep trench in the wet land you would find water and you wouldn't be able to plant in water.'

'Well spoken,' he said. 'When the trenches have been dug, do you know at what season it is necessary to set out the cuttings (*phuta*) for each type of land?'

'Yes indeed,' I said.

'Well, wanting them to grow as quickly as possible, do you think that scattering worked earth underneath, the shoot of the cut vine-branch will move down faster through the soft earth or through uncultivated soil into hard ground?' he said.

'Clearly,' I said, 'it would shoot faster through cultivated earth than through unworked.'

'Surely then, the earth should be made into a bedding (*hypoblêtea*) for the cutting (*phutos*).'

'Certainly,' I said.

'Will setting the whole cut vine-branch upright looking towards the sky lead it to root better, or at an angle, placing some under the earth bedding, so that it lies on its side like the letter gamma?'

'That way, at an angle, by Zeus, for there would be more buds (lit. 'eyes', *ophthalmoi*) underground, and from the buds above ground the cuttings (*phuta*) produce shoots. And, I think the buds below ground also do the same things. When there are many shoots growing from the ground I think the cutting (*phuton*) will shoot quickly and strongly.'

'Also about these things,' he said, 'you know the same things as I do. And would you just heap up (*epamêsaio*) the earth,' he said, 'or would you pack the earth around the cutting?'

'I would pack it, by Zeus,' I said, 'for I know well that water would turn unpacked earth to mud, and it would be dried by the sun to some depth, so that the cuttings (*phuta*) would be in danger of rotting from being too wet or withering from being too dry when the roots were too hot.'

'About planting vines then, Sokrates, you seem to know all the same things that I do.' (Xen. *Oec.* 19.3–12.)

Rooted cuttings were cuttings either taken from layers (branch wood set in soil to root while still attached to the tree), or else set to grow and produce roots in a nursery bed before they were planted in their final position (Fig. 5.1a). There is considerable evidence for the practice of establishing nursery beds in the Roman agricultural treatises, and Theophrastos (*CP* 3.5.3) implies their existence in classical Greece. Certainly, he highly recommends the use of rooted cuttings (Theophr. *HP* 2.2.4). But there is almost no direct evidence for the actual use of nursery beds for cuttings anywhere in classical Greece, and Theophrastos (*HP* 2.5.3; *CP* 3.2.7; 3.3.1; 3.4.2; 3.5.3) makes it clear that unrooted cuttings were widely used. In the Attic Stelae, in other inscriptions of the *poletai*, and even in the leases which specify that trees must be planted out[2], there are no mentions of nursery beds. Examples of well-documented propagation practices are admittedly few, but the limited evidence suggests that the technique of planting cuttings first in nursery beds for several years before transplanting to the final location was rarely (if ever) used in classical Greece. One reason for this is likely to have been the irregularity and small sized areas of tree planting, as explained in more detail below (section 5.8).

5.3. WATERING CUTTINGS

Once cuttings are planted out they need an enormous amount of watering through the summer for several years, if the failure rate is to be kept to a minimum (Theophr. *CP* 3.8.3). How much depends partly on when they are planted. Once cuttings (or older trees) are

[2] E.g. Rhodes and Osborne 2003: 282–6, no. 59. *SEG*[3] 963, from Amorgos; cf. Osborne 1987: 37, 42–3, Foxhall 1996.

used to being watered in hot weather, they can rapidly perish from water stress if the level of watering to which they are accustomed is not maintained (Theophr. *CP* 3.8.4). Theophrastos (*CP* 3.2.6) specifies that, just as today (Forbes 1982: 232–3), tree planting was usually done in either spring or autumn. Theophrastos (*CP* 3.3.1–2) preferred the spring planting season, because the cuttings have a chance to become well established before the onset of the cold weather. But he obviously assumed that farmers would have had plenty of labour available (or normally planted close to home) for if cuttings are planted in the autumn, they need less, and less frequent, watering, despite the fact that more would probably succumb to cold (Pansiot and Rebour 1961: 89). I have no precise figures for the amount of water and frequency of watering necessary for unrooted cuttings, but it is likely to be considerably higher than the figures given below for the watering of young trees planted out with roots.

5.4. PLANTING HOLES AND TRENCHES FOR CUTTINGS

Theophrastos (*CP* 3.4.1; *HP* 2.5.1) recommends that the holes (*guroi*) for planting trees be dug a year or so ahead of the intended planting time. Modern practice also recommends that planting holes be dug several months before a young tree is put in place (Pansiot and Rebour 1961: 89). The recommended depth of a maximum of three Greek feet (about 0.89 m) is in line with modern recommendations, though the exact depth of planting depended on the soil of the particular location and the size and development of the young tree or cutting planted out (Xen. *Oec.* 19.2–5; cf. Theophr. *HP* 2.5.1–2; Pansiot and Rebour 1961: 88). Soil in the hole was broken up, and Theophrastos (*CP* 3.4.3) even mentions a number of special treatments and inclusions such as brushwood or vine prunings, chunks of wood or pots of water. This suggestion seems in line with the modern recommendation that slow-decomposing organic matter be placed in planting holes (Pansiot and Rebour 1961: 87). Cuttings can be planted vertically or horizontally (Pansiot and Rebour 1961: 69, fig. 22). Xenophon (*Oec.* 19.9) describes what is probably a kind of horizontal cutting, and Theophrastos (*CP* 3.5.4) mentions both. Once in place, cuttings were almost completely buried, to encourage maximum root growth (Xen. Oec. 19.9.9–10; Theophr. *CP* 3.5.4; cf. *CP* 3.4.2; *HP* 2.5.4; Pansiot and Rebour 1961: 99). This is all the more important for unrooted cuttings. Vertical cuttings were cut right back (Theophr. *CP* 3.5.4–5) and on olive cuttings in particular the cut was sealed to prevent water damage. However, the hole was not entirely filled in so that a hollow basin (the *guros*) was left around the buried cutting (Theophr. *CP* 3.4.2). This caught and retained rain water and irrigation water where it would most benefit the developing tree. Or conversely, in damp areas and/or in wet winters, the basin could be connected to a system of ditches to control runoff (see below, section 5.10).

5.5. LAYERS AND SUCKERS

Layers were treated essentially as rooted cuttings (Theophr. *CP* 3.5.1), which is more or less what they are (Fig. 5.1e). The technique, known as layering, consists of placing a thin branch of the tree partially under the soil so that it can develop roots while it is still attached to the parent tree. The technique is not usually used with olives, in large part because on most olive trees the newer growth is too high for the technique of burying young branches in the earth to be practicable. It is, however, a very important propagation method for vines, both in antiquity and at present in Greece and elsewhere (Theophr. *CP* 3.11.5; Forbes 1982: 256–8; Hanson 1992: 163). There is archaeological evidence in the form of buried plant pots probably used for layering that the technique was in use for vines and other plants in the sanctuary gardens of the Hephaisteion in Athens (D. Thompson 1937: 419). The main advantage of layering is the high rate at which the plants take root: this is particularly important in the case of the vine, which does not root so easily as the olive or the fig from simple branch cuttings, and which is much more difficult to graft.

The simplest means of producing a rooted stock for olives, equivalent to planting out a layer, is to pull off and plant the suckers which spring from the base of the tree. There are two difficulties with this method, however. First, suckers are normally removed before they are large enough to plant out on well-tended olive trees, so the technique is really only useful for rejuvenating a neglected grove. Second, if trees have been grafted, suckers will, of course, only reproduce the root stock, which is often an inferior fruiting or 'wild' cultigen (e.g. the oleaster). Although the technique is mentioned in the ancient sources, there is little evidence to suggest that it was widely used. Both layers and suckers come into production faster than cuttings.

5.6. OVULES

The best documented technique for propagating olives in classical Greece was the planting of ovules (Fig. 5.1d), a technique which has been almost unrecognized in ancient sources:

> Some trees also have what are called by some 'growths' (or something corresponding), such as the olive; for the name belongs most properly to that tree, and it seems inclined to develop them. Some call this a '*premnon*', some a '*krotone*', and others a different name. (Theophr. *HP* 1.8.6).

Elsewhere when he refers to ovules, Theophrastos (*HP* 2.1.4; 2.5.4; 2.5.5) normally uses the term '*premnon*'. There are many advantages to planting out ovules (Pansiot and Rebour 1961: 74–5). First and foremost, they are highly drought-resistant, which considerably reduces, or in some cases, especially with autumn planted trees, even eliminates the need for watering. Concomitantly, rooting is easier, the success rate high, and trees develop comparatively quickly. The disadvantages are; 1) their limited availability, as they must be removed from mature trees; and 2) the removal of too many from the parent trees can weaken, or, in extreme cases, even kill the parent. The combination of their limited availability and great desirability must have made them a highly marketable commodity.

Once ovules are removed from the parent tree they rapidly begin to sprout shoots if they are left for any length of time, without being allowed to dry out. Such growth can be encouraged if they are heeled in, and they can then be planted out on the permanent site after shoot growth has commenced, either before or after roots have appeared (Pansiot and Rebour 1961: 75–6). This is certainly the method of propagation described by Xenophon (*Oec.* 19.13): 'You see that ovules adhere to all the young trees for planting out, and you see', he said, 'that mud is smeared on all the tops of the trees, and the upper part of all the trees is roofed over' (Cf. Amouretti 1986: 59).

Here the ovules have been left to shoot, and after planting out, the shoots have been cut hard back, coated with mud, and a pot sherd placed on top (Xen. *Oec.* 19.14) to keep out moisture which would rot the young tree. The ovule may also have been coated with mud when it had been cut, before it was planted, to keep it from drying out (cf. Pansiot and Rebour 1961: 75). It is not clear from this passage whether or not they were actually rooted, or had been grown first in a special nursery bed. Both are possibilities, but the passage itself does not specify. Theophrastos (*HP* 2.1.4; 2.5.5) describes the use of ovules for planting out olive trees similarly. His account implies that quite large ones were cut away from the tree and planted on the permanent site almost immediately. They were laid in the earth cut side down, and soil was heaped up around them as the shoots developed.

However, in order to exploit the potential advantages of propagation by ovules, large ones, weighing from 1.5 to 5 kg, must be used. The technique consists of making a saw cut on either side of the ovule, then adzing out the body of the ovule between the two saw cuts (Pansiot and Rebour 1961: 74; cf. Theophr. *HP* 2.1.4; 2.5.5). As noted, the removal of such large ovules can severely weaken the parent trees. This is certainly what is going on in Demosthenes 43.68–72:

> One thing they have brought about most contravenes the law and is most filthy, and reveals that they care for nothing except profit. For no sooner was Theopompos awarded the estate of Hagnias in the way you have heard, than immediately he showed that he thought it didn't belong to him. For what was of greatest value in Hagnias' fields and most wondered at by those having neighbouring plots and by all other people was the olive trees. They have dug out and cut out the ovules on these, more than a thousand stocks, from which much oil came. Having removed the ovules from them they sold

them, and made vast amounts of money. And they did this while the estate of Hagnias was still subject to adjudication according to the law which allowed them to bring forward the mother of this boy.

To show that I speak the truth in this and that they removed the ovules from the olives from the fields which Hagnias left, I will produce witnesses for you of these things, both the holders of neighbouring plots and others whom we called forward, when we appealed for witnesses of these things. Read the testimony.

Testimony

They testify that, having been summoned by Sositheos, they followed him to Hagnias' land at Araphen, when the estate of Hagnias had been judged to be Theopompos', and Sositheos showed them the olive trees having had ovules removed from Hagnias' land.

If, men of the jury, they had violated only the dead man in doing this, they would have done a terrible thing, but less bad than what they have done now that they have violated the whole city and contravened the law. You will understand when you hear the law. Cite the law.

Law

If someone should dig up an olive tree in Athens, unless it is in a public sanctuary of the Athenians or one belonging to a deme, or up to two olive trees per year for his own use, or it would be necessary for the use of the dead, he shall owe 100 drachmas to the public treasury for each olive tree, and one tenth of this shall be for the goddess. And he shall owe 100 drachmas for each olive tree to the private citizen prosecuting him. Cases concerning these matters are to come before the archons, of which things each and every one are judges. Let the prosecutor deposit his share of the court fees. When someone is found guilty, let the archons before whom the case came register it with the officials, the amount due to the public treasury, <and the amount due to the goddess,> with the treasurers of the goddess. If they don't register it they themselves will owe it.

The law is thus strong. But consider this, men of the jury, what you think we have suffered from these men in times gone by, and by the outrageous behaviour these men have shown to you, so great a city, for they have been contemptuous of your laws, and have done what the laws specifically forbade, having thus contemptuously maltreated the fields which Hagnias left. For the law forbids doing these things on one's own patrimonial land. But much do they care for your laws, or that the house of Hagnias not be deserted!

The speaker, Sositheos, could be attempting to mislead the jury by citing a law which is not in fact relevant to the offence he claims

Theopompos has committed, if the law belongs in the text.[3] Only once are the olive trees said to have actually been 'dug out' (Dem. 43.69): ἐξώρθττον καὶ ἐξεπρέμνιζον, and even here the verb is accompanied by the one more regularly used in the passage for Theopompos' action. The witnesses only testify that ἐξεπρέμνισαν (Dem. 43.70). Never are words used which unambiguously denote digging up or cutting down trees. ἐκπρεμνὶζω and πρεμνὶζω are difficult words, found only in this passage and the testimonia to it. They are explained as 'root out' by the late lexical writers, but there is no evidence to support this interpretation. Most likely, this meaning is derived by later (including modern) commentators from the cited law (Dem. 43.71), which really does refer to the penalties for digging out and uprooting olive trees.

The care with which the speaker chose his terminology is not accidental. The word πρέμνον has several meanings, but in a passage relating to olive trees it is most likely to be an ovule. πρεμνίζω and ἐκπρεμνίζω, then, must surely mean 'to cut out ovules', not 'to uproot' or 'to dig out'. Since the technique of removing ovules used an adze, and often they are located very close to the base of the tree, the verb ἐξορύττω might just about be applicable to this action (especially since obfuscation was intended by the speaker: if jurors misunderstood and thought that trees had been dug out this would only enhance the chances of a successful prosecution). Theophrastos (*HP* 2.1.4; 2.5.5) in his brief descriptions of planting olives from ovules in fact uses the verbs κατακόπτω and διακόπτω, 'cut off, cut through' to describe their removal from the parent tree. It is noteworthy that the law cited (Dem. 43.71), *only* uses the word ἐξορύττω, not πρεμνίζω or ἐκπρεμνίζω. The speaker's one use of ἐζορύττω may therefore be an attempt to show that the offence committed by Theopompos genuinely contravened the law he quotes, which in fact it did not. After the law is quoted, the speaker refers to Theopompos's actions in relation to the law only in vague terms: 'the law forbids that anyone do these things, even on his patrimonial land'

[3] It is generally believed that the laws 'quoted' like this one in Athenian law court speeches were inserted by later copyists. The status of these laws is uncertain; some may be genuine laws, though not necessarily the ones cited by the speakers originally, whilst others may be forgeries.

(Dem. 43.72). Moreover, there is no indication that Theopompos was prosecuted under the actual terms of the cited law. Presumably, Sositheos reckoned he did not have a good enough case for such a prosecution to be successful.

What Theopompos has almost certainly done is to remove large numbers of ovules from the trunks of Hagnias' mature olive trees. This would certainly have reduced their productivity, and may well have damaged or killed some trees. These ovules could have been used to start new olive trees on other land more securely owned by Theopompos, or else sold: big ovules would have been a commodity in limited supply since farmers would have been well aware of the damage their removal might cause to productive trees. Essentially he is 'asset stripping', but however despicable the action might have been considered, there is no evidence that what Theopompos did was actually illegal. The other, less likely, possibility is that Theopompos was rejuvenating olive trees that had been neglected. Given the amount of time that Hagnias' land seems to have been disputed, this is just about possible. Certainly this would have been the obvious defence for Theopompos: that he was simply removing extra trees and suckers in an overcrowded stand and carrying out heavy pruning in order to restore productivity. But the first alternative, that he was removing ovules for sale (as Sositheos accuses) or for his own use elsewhere, seems the most likely.

The speaker in Lysias 7 defends himself against a similar charge in the latter part of the speech. Here, the initial accusation had been that the speaker had removed a sacred olive tree. This was changed by the prosecution to a charge of removing an enclosure where there had once been a sacred olive tree (see below, section 5.9). Most of the speech concerns this second charge, but at Lys. 7.19, the speaker reverts to defending himself against the first charge and claims to repeat his accuser's account of how he had his slaves cut out the ovules, then had the wood loaded onto a cart when he removed a sacred olive: 'He says that I stood by, and the slaves cut out the ovules, while the ox-driver loaded up the wood and hauled it off'.[4]

[4] Cf. Lys. 7.22, where it is made clear that this passage refers to the accusation of removing the sacred olive tree (*moria*): 'and even if you say you saw me removing a sacred olive tree . . .'.

The passage is plainly phrased to focus on the two of the most profitable aspects of destroying an olive tree; it is not intended to describe the process of its removal.

5.7. GRAFTED TREES

Grafting was a well-known technique in ancient Greek agriculture (Fig. 5.1b–c).[5] Theophrastos (*CP* 1.6; cf. *HP* 2.1.4; *CP* 1.6.10) includes a long discussion of grafting in general, and specifically notes the practice of grafting cultivated olives onto 'wild' stocks, once the latter have been established in a permanent location. He describes both crown grafting, where the stock is cut back and twigs are inserted vertically behind the bark, then the top is sealed to prevent the entry of water and infection, and bud grafting, where a patch of bark is removed, and a patch containing a bud of the desired cultivar is inserted and bound tightly to the tree in its place, then sealed. Both techniques must have been used on olives, for the olive accepts both types of grafts easily. Of all the fruit trees grown in classical Greece, the vine is the most difficult to reproduce by grafting with the highest failure rate of grafts (Theophr. *HP* 2.5.3). Figs, like olives, were easily grafted. The advantages of grafting were that specific (and especially desirable) cultivars could be easily and quickly reproduced, and that risks of sensitive or non-local cultivars could be reduced by grafting onto a vigorous stock. It is interesting that one of the few farms for which there is secure evidence of relatively large-scale arboriculture, including olive cultivation, clearly used olive trees grafted onto wild

[5] How far back the arts of grafting and budding go in Greece is uncertain. These techniques were in use in the ancient Near East in the Bronze Age, and there are technical terms associated with grafting (notably *arkat kiri*, literally, 'inheritance of the orchard', for a shoot or scion) attested back as far as the Old Babylonian period (Powell 1987). The technique is likely to have developed in Greece at an early date: my preference is for a Bronze Age date, but the evidence is inconclusive. Hanson (1999: 42–3, 78, 81) maintains that grafting and budding were part of a later Dark Age 'revolution' in agronomic practices, which included the introduction of arboriculture on a substantial scale, associated with the evolution of the polis. In fact, what little evidence there is suggests that the practices and techniques of arboriculture were maintained from the LBA through to the Archaic period (Foxhall 1995), and they are likely to have been introduced and regularly used much earlier than Hanson asserts.

Fig. 5.2. Bud-grafted 'wild' olive stock on Methana, 1984 (L. Foxhall).

stock. The 'Thaleion' farm on Mykonos, owned by the Temple of Apollo on Delos and worked under lease contracts, had 147 cultivated olive trees, 87 wild olives that had been grafted, 200 wild olive trees, as well as 1140 vines, 143 fig trees, and 101 apple trees (Kent 1947: 288).

In modern Methana, grafting domesticated cultivars on to 'wild' stock has been the normal propagation method for both olives and pears (Fig. 5.2) (Forbes 1982: 254–5). For olives, young 'wild' olive trees are dug up in the mountains and transplanted onto the permanent site, usually around February. They are watered for at least two summers, how often, how much, and how long, depending on the moisture retentiveness of the soils, the spacing of trees, the exposure, and the amount of rainfall. During the first summer young trees need at least five to six waterings (about every ten days to one month over the summer/early autumn period). The total water consumption for the first summer is around 150–250 litres per tree. During the second summer young trees are watered two or three times, using perhaps 50 litres of water per watering (100–150 litres altogether).

Fig. 5.3. Young olive tree planted in basin/trench, with brushwood covering for protection and to reduce evaporation, Methana 1984 (L. Foxhall).

Since olives (and many other crop trees) are often planted in inaccessible places, distant from wells, cisterns and other water sources, the water for young trees must be carried by donkey back to them. Like cuttings, young stocks and grafted trees also need to be protected against sun scorch, so trunks are wrapped or whitewashed. Frequently, young trees are protected from the double ravages of livestock and the summer sun with wrappings of spiny broom. If labour is available, young trees are planted in basins, which contain the water poured on them, and directs it to the roots. Sometimes these are covered over with brushwood to help cut down evaporation (see Fig. 5.3). The failure rate of young olive trees is quite high, but the labour needed for more and longer watering of young trees is not available in peasant households. In the southern Argolid, with less moisture retentive soils, young olives are watered for up to five years.[6]

[6] I am grateful to Hamish Forbes for much of this information.

This is where wealthy households in classical antiquity, who had slave labour available, would have been at a considerable advantage in establishing young trees. A slave labour force might have had time to water intensively during the summer, thus reducing the failure rate of young trees and possibly promoting faster growth. Trees are not generally grafted until the stocks are well established, now or in antiquity (Theophr. *CP* 1.6.10).

5.8. LOCATIONS AND SPACING OF OLIVES AND OTHER CROP TREES

Undoubtedly, crop trees were often grown in locations and in soils that were far from optimal. As Theophrastos (*CP* 3.7.5; cf. 3.11.1–16.4) realised, vines are most exacting about soil and exposure requirements, and even they are tolerant of a wide range of conditions. Selective breeding expands the tolerance range, since some cultivars of any particular cultigen can usually be persuaded to grow even in apparently adverse conditions.[7] Almonds, figs, and olives do better on rich, deep, soil than on poorer soil, but will bear adequately on thinner, less fertile soils on slopes (Theophr. *CP* 3.6.7–8; *HP* 3.2.5). In any case, the very best land must often have been reserved for cereals, and trees were relegated to second quality land (Theophr. *CP* 1.18.1–2). But the olive in particular is renowned for its ability to produce on poor soils and steep slopes, as long as they are not at too high an altitude. On modern Methana, the altitude limit for viable olive cultivation is around 300–400 m above sea level, as trees will hardly grow above 500 m. Figs, almonds, and even vines thrive at much higher altitudes, with vines doing well as high as 750 m, sometimes in conjunction with apples, plums, and pears. In Southern Calabria, however, olives appear to thrive at altitudes up to about 600–650 m above sea level, but not usually higher. Exposures are as important as altitude for tree growth, those in southern exposures coming on better and more rapidly than those facing north. (Theophrastos *CP* 2.3.1–3; 2.4.8; 3.5.2; 3.6.9; 3.7.9; Pansiot and Rebour 1961: 44). The practical altitudinal limit for olive

[7] For example, the water-loving Lakonian fig, discussed by Theophrastos, *CP* 3.6.6; *HP* 2.7.1, or the nineteenth-century variety of fig, 'Brown Turkey', which will fruit even in British summers.

cultivation does not seem to be the same throughout Greece, let alone throughout the Mediterranean region: there are many local factors which influence how high and how far inland olives can be grown.

On the whole, it seems that wealthy farmers in classical Greece did not usually engage in the polycropping of arable and tree crops which has been a regular feature of peasant cultivation regimes in many parts of modern Greece through the twentieth century, or at least there is little or no evidence for the practice. Theophrastos (*CP* 3.10.3; 3.15.4) considered it a useful technique only for damp areas, and Xenophon does not mention it at all, nor does it appear among the cultivation regimes recommended on lease documents. In Lysias 7.7, the speaker claims that when many Attic olive trees were cut down early in the fourth century the land became arable (see section 5.9). This statement, even if untrue, suggests that many Athenians might have thought of tree crops and cereals growing on different plots. One reason for this may well have been the potential for increased erosion that could have occurred on sloping but un-terraced fields, or the difficulties of trenching trees in fields also planted in arable crops (see section 5.10). Other possible reasons are that it may not have been as effective a cropping strategy for using slave labour, or that for wealthy farmers land was not in short supply. On the other hand there is considerable evidence for planting several different kinds of crop trees on the same plot, a cultivation regime that is also in use now in many parts of Greece and the Mediterranean region. The utility of this system is that short-lived trees such as almonds (or nowadays citrus) can be interplanted with long-lived but slow growing trees such as olives, bringing a return on the plot earlier than if it were planted in olives alone. By the time the olives come into full production and are very large, the short-lived trees can be removed. Theophrastos (*CP* 3.10.4–19) discusses the benefits and disadvantages of various combinations of tree crops in some detail.

This is not to say that big landowners never planted arable crops and trees in combination, or that small-scale farmers always did. But what little evidence we have suggests that large-scale farmers generally preferred not to use this technique. There are, in fact, two possible examples of tree crop/arable polycropping in the Attic Stelae. In one passage (*IG* I^3 422.81–6) figs, grapes, and olives are sold along with chaff and bean-straw which are described as 'on the same land'. This could imply polycropping, but it could mean that the crops were adjacent to each other or that the tree crops were planted around the perimeter of the plot. The other passage (*IG* I^3 427.72–6) refers first to 'arable land' (*gê psilê*), but goes on to mention vines on it. Pritchett (1956: 263) inferred from this passage that the term *gê psilê* meant land with cereals or vines on it, but not fruit or olive trees. This is unlikely to be correct, because it is clear that most classical sources classed vines as trees. What this is most likely to refer to is a piece of land with both vines and arable crops on it. The most likely arrangement is that the vines were planted along one edge of the plot. Using this interpretation, is easy to restore ϵ[.....] in line 75 to read *elaôn*, thus making lines 74–5 read as: [ἕτ]ϵ[ρος]ἅγρος γε̂ς φσιλε̂ς πλέθ[ρα] | ἐ[λαο̂ν]||||, 'another piece of arable land, with olives, 3 plethra'. There is considerable evidence for the practice of planting olives around the outside of a plot (see below and Chapter 7.3) and it seems most sensible to understand such an arrangement here. Olives, then, were certainly planted in plains land, but normally together with other tree crops. However, they must also have been planted in areas of steep slopes and thin soils where little else would thrive.

Aside from the examples of olives planted with other tree crops on leases (Osborne 1987: 42–3), the Attic Stelae and a few literary references provide the evidence for the locations of olive trees. Only one example, the grove of olives on Hagnias' fields, even suggests the monoculture of olives, and it is clear that these fields do not represent the whole of Hagnias' landed property (Dem. 43.68–72; cf. Is. 11). In contrast, Apollodoros' ornamental plot which was vandalized by his neighbour and (possibly) client ([Dem.] 53.15, quoted in Chapter 7.3), was planted in table fruits, most likely (though not necessarily) figs bred specially to be used as fresh figs (cf. Pl. *Leg.* 8.844d–e), which had vines trained to climb up them (Theophr. *CP* 3.10.8; 5.5.4), and young olives were planted in a line around the circumference of the plot. Although the context is comic and therefore not necessarily entirely realistic, the configuration described in *Acharnians* 994–9 (quoted in Chapter 7.3), with a mixture of young figs and vines, enclosed by a row of olives around the circumference of the field sounds remarkably like Apollodoros' plot. Also significant is the joke on the length of time it takes for olives to produce a decent crop. It is clearly meant to be ridiculous that an old man should expect to bathe with oil from olive trees he has only just planted. In both these cases it is likely that the planting scheme of these plots aimed to be decorative as well as economically viable.

On the Attic Stelae, there are ten instances when crops in the field are sold (*IG* I^3 421.20, 22, 29; 422.81–6, 87–9, 97, 218; 428.5; 430.34). Where the context is complete enough to be sure that the crop was not specified, it is most likely cereals should be assumed. But in two instances the produce is specified as 'figs, grapes and olives' (*IG* I^3 422.81–6, 87–9). Clearly, the trees were at least in close proximity, if not actually on the same plot.

The impression that trees, especially olive trees, were often planted on the edges of plots emerges from other sources as well. A passage in the Attic Stelae with two probable examples of this arrangement appears at *IG* I^3 427.72–5. In Demosthenes 55.13 both vines and figs were planted on the plot belonging to the speaker, and there were family tombs on the land as well. The dispute in this speech centred on damage caused by run-off, allegedly because the defendant (the speaker) had walled off the outlet for the water. A close examination of the text (especially 55.10) suggests that the disputants did not have plots one above the other (as is normally construed) but on either side of a large gully, which at some parts of its course coincided with a road. The speaker's trees were probably planted along the road, like those he alleges his opponent to have (55.22), which were planted along the road and then walled in when the trees were quite large (Foxhall 1996). As regards olives, Xenophon explicitly says that they were frequently planted along the roadside (Xen. *Oec.* 19.13).

The implications of this for the spacing and planting configurations of crop trees are interesting. Olives, figs and other large crop trees were sometimes planted in grids, in a quincunx arrangement, or even in rows, though monocropping may have been rare; more usually several tree species shared a plot. However, such planting patterns are likely to have been restricted to relatively large areas of flat or gently sloping land, and were almost certainly confined to the estates of the wealthy. Part of the cachet of being able to devote large areas of such prime land to trees was in many cases the potential to plant it in a decorative way, particularly in the case of roadside plots where it was visible to passers-by (see Chapter 7). The olives and figs of poorer farmers may have been more randomly positioned on their plots, as was governed by considerations of size and space available, slope and the presence of established 'wild' trees suitable for use as rootstocks for grafting. In fact, on more steeply sloping land, it is more beneficial to plant trees in accordance with the contour, rather than to a preset pattern (Pansiot and Rebour 1961: 92–3, 95), and this is even more true when fields are not terraced. With vines, however, which are the tree crop most often planted on flat land, and for which the spacing is closer than for large trees, arrangements of plants in regular rows seem to have been more common, and perhaps the use of regular planting patterns was spread more widely across the socio-economic spectrum (Xen. *Oec.* 20.3–4; Theophr. *HP* 4.4.8).

Optimal spacing for olives, figs, and other large trees varies with the quality of the land and the amount of moisture available. Generally, on poorer land trees were closer together than on good land (Theophr. *CP* 3.7.2; *HP* 2.5.6; Pansiot and Rebour 1961: 90–1). On rich deep soils and/or with an abundance of water available to the trees, they grow bigger. If they are too closely planted branches crowd each other out and roots become entangled. For olives, which have many wide spreading roots near the surface, this is particularly problematic. Conversely, on poorer, shallower soils, especially in the absence of moisture, trees remain smaller and can thus be planted closer together. In areas of very low rainfall, however, especially <200–300 mm, trees must again be widely spaced in order that they do not over-compete for soil moisture. This does not affect most of Greece, and is more important in places like North Africa.

The spacing of trees then (at least those of successful growers), is heavily determined by biological and environmental factors. It is therefore not surprising that the spacing recommended in classical antiquity was almost exactly the same as that used in Greece today. The modern spacing of Methana of around 10 m is relatively dense, allowing a maximum of 100 trees per ha. This is in accord with Theophrastos' recommendations on spacing (Theophr. *HP* 2.5.6). Similarly, in a law attributed to Solon (Plut. *Sol.* 23), olives and figs, both of which have wide spreading roots as mature trees and cast a great deal of shade, should not be planted closer than 9 Greek feet (c. 2.7 m) to a neighbour's plot, is almost exactly the same as the modern Greek law which specifies the distance as 3 m. Amouretti's (1986: 26) proposed maximum planting densities of 200–250 trees per ha on good soils in areas of high rainfall therefore appear unrealistic in light of well-documented ancient and modern practice in Greece.

5.9. THE SACRED OLIVES OF ATTICA

In Attica there is the special problem of the sacred olives (*moriai*, sing. *moria*). We know very little about these trees which were located mostly on private land, but subject to state control and sacred to the goddess Athena, to whom a proportion of the produce was paid. The main texts are Lysias 7 and a short passage of the Aristotelian *Constitution of the Athenians* (60.1–2). The latter passage is worth quoting in full:

They also allot ten men as *athlothetai* [magistrates in charge of the prizes], one from each tribe. These men, having been scrutinised, hold office for four years. They administer the procession of the Panathenaia, the music contest, the athletic contest, and the horse race; also they have the *peplos* [the dress for the statue of Athena] made, and with the Council they have the amphoras made, and they distribute oil to the athletes. The oil is collected from the sacred olive trees, for the Archon places a levy on those owning plots on which there are sacred olives, three half-kotylai from each trunk. Formerly the city used to sell the fruit, and if someone would either dig out or take down a sacred olive tree, the Council of the Areiopagos judged him, and if it found him in the wrong, they punished him with death. Out of these traditions come the present law that the one owning the land pays the oil, but the judgement is abolished. And (now) the oil for the city is reckoned out of the property, not by the trunks. Therefore the Archon, collecting whatever oil there is for each (property), pays it over to the Treasurers on the Acropolis, and he does not, as formerly, take it up to the Areiopagos until he would hand the whole lot over to the Treasurers. But (now) the Treasurers keep watch over the other year's harvest on the Acropolis, and they measure it out to the *athlothetai* for the Panathenaia, and the *athlothetai* measure it out to the victors in the games. For the prizes for those winning in music are silver and gold (crowns?), for those winning the male beauty contest the prizes are shields, and for those winning the athletic contest and the horse race, oil. ([Aristotle] *Ath.Pol.* 60.1–2).

If we can judge from the amounts of sacred oil given out in prizes in the Great Panathenaia (Shear 2003), recorded on stone in an inscription dated to the first half of the fourth century, there must have been considerable numbers of sacred olive trees. The preserved entries on *IG* I^2 2311 record over 1113 amphorae. The standard painted Panathenaic amphorae used as prizes in the games normally have a capacity of 38–39 litres, suggesting that in the year of this inscription approximately 42,294–43,407 litres of oil was awarded as prizes (Neils

1992: 39).[8] If each tree produced about 3.5–4 kg (3.36–3.84 litres) per major olive harvest,[9] and there were two olive harvests plus small amounts of oil from the 'off years' available as prize oil for each Panathenaia,[10] then the prize oil represented in *IG* I^2 2311 could have been produced by the total crop of about 5500–6500 trees. At 90–120 or so trees per ha. this would represent about 500–700 ha, if they were all planted together (which we know they were not). If the figure given in the *Ath. Pol.* of three half-kotylai (0.405 litre) of oil per tree going to the state for prize oil is correct for the time of this inscription, then there must have been about ten times this number of *moriai*: 55,000–65,000, though this sounds very high. It is interesting, however, and possibly significant, that the figure of the 'tithe' of oil owed to the goddess from the *moriai* would, on the production figures postulated here, represent a genuine if rather approximate 'tithe' of about 10% of each tree's oil crop. This would certainly be in line with the first fruits 'tithes' documented for other cults (Isager and Skydsgaard 1992: 173–4).

The few mentions of sacred olives in the literary record suggest they were scattered randomly(?) across the Attic landscape, though why and when they were planted, and how any particular tree came to be considered 'sacred' in the first place is obscure. The *Ath. Pol.* passage quoted above suggests they were perceived to be very old by the second half of the fourth century BCE. By this time practices had changed several times over and a number of rather confused traditions circulated about how the city 'used to' manage the sacred olives (e.g. the statement that the city 'used to sell the fruit' seems completely inconsistent with the other practices described as 'from past times'). However, Lysias 7, dated to sometime after 397 BCE when the alleged offence was committed, not so very many years before the *Ath. Pol.*, seems to be a case tried by the Areiopagos (the court is addressed as '*boule*' throughout), following one of the practices mentioned as outdated

[8] On Panathenaic amphorae see also Amyx 1958: 178–86; Boardman 1974: 167–77. Among dated Panathenaic prize amphorae none has a date coinciding with a festival year, suggesting that oil was stored in them as it was produced and placed under the guardianship of Athenian magistrates, Boardman 1974: 169.

[9] For olive oil, 1 litre = 0.96 kg. I have used a higher 'average' production figure here than in Chapter 6, on the grounds that *moriai* in the fourth century were probably mostly large, old, trees which cropped quite heavily compared with younger ones.

[10] Significantly, when [Arist.] *Ath. Pol.* describes the past and present procedures for storing oil on the Acropolis, the stored oil is called 'the other year's harvest' (the passage is quoted above), suggesting that there really were two substantial oil crops within the four year period between major Panathenaic festivals.

by the *Ath. Pol.* In Lysias 7.2 the speaker specifically mentions 'those who bought the fruit of the sacred olives', another practice relegated to the past by the *Ath. Pol.* Further, in Lysias 7.25 and 29, the speaker makes a point of saying that the care and administration of sacred olives was the responsibility of the Areiopagos, while *Ath Pol.* indicates that the archons largely carried out these duties. Was there some radical change in the administration of the sacred olives during the fourth century, or was the writer of the *Ath. Pol.* simply confused? It is unlikely that the trees were planted in regular patterns: Lysias 7.14 envisaged a scenario in which the disputed sacred olive enclosure might have been thought to be impeding vines or close to a building:

> 'However this man could not show either that I was compelled by poverty to attempt such deeds, or that the plot was spoilt for me by the presence of the enclosure, or that it was an impediment to vines, or that it was near a structure, or that I was unaware of the dangers at your hands.'
>
> (Lys. 7.14.)

Would the presence of these trees in significant numbers have affected the planting of other trees and arable crops since they could not be tampered with? It is not clear either how they were distributed amongst landowners.

The speaker of Lysias 7 claims that many plots had been 'thickly wooded' with private and sacred olive trees before the war and these have now been cut down and these areas have become arable land, but given the forensic context, this is very likely to be an exaggeration which should not be taken a face value. 'You are aware, O Council, especially as many of you who are in charge of such matters, many plots in that time were thickly wooded with private and sacred olive trees, of which now the majority have been cut down, and it has become arable land...' (Lys. 7.7).

Elsewhere the speaker notes that among his plots on the plain he has a number of 'private' olive trees as well as sacred olives (Lys. 7.24). Did most grow on the land of the wealthy, or were they more evenly distributed across the socio-economic spectrum? For how many might any one landowner have had to take responsibility? One interesting feature mentioned in the *Ath. Pol.* passage quoted above is that 'traditionally' the levy of sacred oil had been collected 'by the trunk' (*stelechos*). This provision, using the semi-technical term *stelechos*, rather than the more normal, everyday word for 'tree', *phuton*, sounds almost like a quotation from a law, couched in language intended to prevent at least some arguments with landowners over the collection. It might also suggest that some of the *moriai* were very old, to the point that they had fallen down and regenerated into multi-stemmed trees.

What sacred olive trees looked like is reasonably clear thanks to Lysias 7. This speech was part of the defence against the accusation that a *sêkos*, 'enclosure', where a sacred olive tree had once been planted, was removed by the landowner. This case was brought to court by the prosecution after a previous suit, alleging that the defendant had destroyed an actual sacred olive tree, had failed.[11]

'Thus the case set against me is puzzling, since first I was indicted for clearing an olive tree from the land, and they went and made enquiries of those who bought the fruit of the sacred olive trees. But when by this means they were able to discover nothing unjust that I had done, now they claim I have cleared an enclosure (*sêkos*), thinking that for me this will be the most difficult charge to refute, while for them it permits them more easily to say what ever they wish.'

(Lys. 7.2.)

The primary meaning of *sêkos* is a sheepfold or similar rustic enclosure (Hom. *Od.* 9.219, 227, 319, 439, 10.412; *Il.*18.589; Hes. *WD* 787). However, it also comes to mean a sacred enclosure, for example, surrounding a hero shrine (Htd. 4.62; Soph., *Philoc.* 1328; Plut., *Kimon* 8; *SIG* 247 K^1 II 55, Delphi, fourth century BCE; IG IV^2(1).102.29, Epidauros, fourth century BCE), or the enclosure around a tomb (Simon. 4.6). The *moriai*, therefore, appeared symbolically as 'mini-*temenoi*' (sacred precincts) in the fields of Attica, separated from the secular world by a physical boundary. The case in Lysias 7, highly politicized though it was, indicates that the *sêkos* was treated as a sacred precinct even if the tree itself no longer existed.

11 It has generally been assumed that *sêkos* means an 'olive stump', for example as translated by Lamb (1930), in the Loeb edition of Lysias, and in LSJ. This 'one-off' meaning is clearly incorrect, as the use of the word in this speech plainly fits the uses normally found elsewhere. What has confused the issue, I believe, is that in the latter part of the speech, the speaker reverts to refuting the earlier charge that he had removed a sacred *olive tree* (not just its enclosure), probably because it was easier to refute. However, in the passage quoted above, it is patently clear that there were two different charges made: it is not entirely clear whether they were both made in the course of this lawsuit, or (perhaps more likely) constituted two different court cases.

Marked off as they were, they would have been immediately identifiable to both local farmers and invading armies.

As discussed in Chapter 4.5, the sacred associations of the *moriai* and their mythical heritage from the tree donated to the city by Athena are likely to have conferred special qualities on the oil they produced. This must have enhanced the already strong symbolic associations of olive oil.

5.10. CULTIVATION OF CROP TREES: TERRACING AND TRENCHING

The main purpose of cultivating olives, or any other tree crop, is to ensure that the plants receive adequate water and soil nutrients. The primary limiting variable for tree growth under Mediterranean Greek conditions is normally the availability of water. According to Theophrastos (*CP* 3.6.2, 3.9.1, 3.9.5), the over-concentration of soil nutrients, from the addition of too much manure or other fertilizers, in the absence of adequate moisture, was likely to do more harm than good, especially to young trees. Although he perceived this to be a problem, given the limited availability of manure and the high rate of decay of organic matter, it probably did not often cause difficulties; at worst, over-concentrations of nutrients will simply not have been available to the trees (Pansiot and Rebour 1961: 118, 110–11, 99). The primary aim of the cultivation practices used by wealthy farmers who could afford the requisite inputs of labour, was to maximize the amount of water (mostly from winter rainfall) available to the roots of the trees. To the same end, repeated digging destroyed roots close to the surface, encouraging the development of roots at lower levels where more moisture was available and where they were less likely to suffer from sun scorch or exposure, hence allowing trees to utilize available water more efficiently (Theophr. *CP* 5.9.8). This was particularly important for olives, which mostly have wide spreading roots close to the surface (Theophr. *HP* 1.6.4; Pansiot and Rebour 1961: 14–16). Basically, this meant lots of digging (Theophr. *HP* 2.7.5; *CP* 3.10.1;3.12.1; 3.20.7).

It is clear that digging was the most laborious and time-consuming part of tree cultivation, for olives, figs, and especially vines. As has already been shown, trees were planted in basins or trenches, which served to catch and hold water close to the base of the tree where it would be available to the roots (Pansiot and Rebour 1961: 176–8). The extent to which this technique was specifically adapted for Mediterranean environmental conditions is revealed by Theophrastos' (*CP* 5.13.1) comments on how it does not work well in regions with cold, wet winters: trees are killed by water standing in the basins and freezing. This further emphasizes the extent to which the technique was normal in the Mediterranean regions of Greece. This increased both the quantity and the quality of the yield. Repeated digging also reduces the weed growth, which would otherwise compete with trees or vines for moisture and soil nutrients (Theophr. *CP* 3.20.9).

How many times trees were dug depended on the local micro-environmental conditions of the plot on which they were growing, the importance of the crop to the owner, and most importantly, the amount of labour available. The outline of digging times given here probably represents a maximum labour input (for example, Theophr. *CP* 3.16.2 suggested that vines be dug three times annually). The soil around each tree was dug several times a year, to shape it appropriately for the season and to loosen it. During the autumn, a basin-like trench was dug out all around the tree. Soil was loosened and banked around the outer edges of the trench to hold water around the base and to allow the winter rains to penetrate to the roots. This also, of course, had the effect of slowing down runoff, and thus reducing erosion as well as keeping runoff water in the places where it would benefit the crop most. In the late winter or early spring, trees were dug again to break up the soil which had become packed down during the winter. While it was still raining in the spring, this further increased the moisture available to the roots and also allowed the sun to warm the soil, thus bringing on budding more rapidly (Theophr. *CP* 3.12.2). Manure or other organic fertilizer was dug into the soil during either the autumn or spring, while there was still sufficient rainfall for the tree to make use of it (Theophr. *CP* 3.9.5; Pansiot and Rebour 1961: 111). Bulky organic fertilizers, in addition to supplying soil nutrients (especially nitrogen), also could increase the moisture retentiveness of the soil. Later in the spring, the top layer of the soil was broken up very finely, trenches were filled in, and the soil heaped up around the bases of the trees. The loose top layer of soil rapidly became completely desiccated by the hot sun. Because soil particles were separated by air pockets, this layer acted as a dust mulch, which prevented evaporation of water via capillary action from the damp soil levels below. Theophrastos (*HP* 2.7.5; *CP* 3.16.3) actually calls the creation and maintenance of a dust mulch 'the dusting'. In the middle of the summer, the soil was broken up yet again (the third digging) to renew and increase the efficacy of the dust mulch. The large scale of these digging operations may also partially explain why the intercropping of arable and tree crops was not recommended in Greece in classical antiquity.[12]

If rich farmers were going to use considerable amounts of labour for digging tree crops three or more times per year in order to increase moisture availability and thus probable yields, then the question arises as to whether or not terracing is worthwhile. The main aims of terracing are:

1) To do something with the big rocks in the soil.
2) To slow down run-off, which means;
    a) that water stays in the same place longer and thus penetrates to deeper levels where it will benefit both trees and (to a lesser extent) arable crops, and is less likely to be lost through evaporation; and
    b) soil erosion is reduced.

All of these aims except 1) are fulfilled by continual and intensive digging and trenching on all but very steep slopes (Pansiot and Rebour 1961: 99, 176–8; Grove and Rackham 2001: 110). Hence, in many areas under the tree cultivation regime of trenches and ditches which was apparently commonly used by large landowners in classical antiquity, there was often no need for terrace walls. Given the substantial proportion of the land in the hands of large landowners (Osborne 1992a; Foxhall 1992), it is likely that this may at least partially explain the rarity of terrace walls in either the archaeological or documentary record which can be securely dated to classical

[12] For the amount of labour these digging operations took under Roman cultivation regimes in Italy see Foxhall 1996: 57–8.

antiquity (see Chapter 3.4.2; Foxhall 1996; Price and Nixon 2005, see Chapter 3.4.2).

It is clear from Theophrastos and other sources that the trenches around trees were only one component in the soil management systems practised by wealthy farmers. Individual trenches around trees were often connected to each other, ideally across the contour to slow run-off further and to prevent the formation of erosion gullies. A failure to carry out this operation properly may well have been part of the cause of the troubles between the disputants in Demosthenes 55 (Dem. 55.10–11, 22, 26–7). In areas of particularly damp soil, on steeper slopes, or where run-off was especially heavy or violent, these small ditches between trees could be connected to larger ditches to drain away run-off without incurring sheet or gully erosion. If there were large stones about that had been removed from the soil during cultivation, these could be profitably placed at the bottoms of ditches, or even occasionally used in the construction of soakaways (Theophr. *CP* 3.6.3–5).

It is evident that this whole system of cultivation, as it is documented in Greek written sources, needed a tremendous amount of continuous labour throughout the year. In classical Greek poleis this was most obviously and effectively provided by slaves. Hence, such cultivation regimes for tree crops must have been the prerogative of the rich. Poor farmers had to find other alternatives, so it may well have been among this sector of the farming community that the building of terrace walls on steep slopes was most frequently practised, to make up for the inability to carry out frequent ditching and trenching.

5.11. PRUNING AND HARVESTING

The pruning (Pansiot and Rebour 1961: 119–23; Theophr. *CP* 3.7.5–12; 3.14.1–15.5) of olives and other fruit trees aims to:

1) encourage the growth of fruit-bearing wood (on the olive this is second-year wood, Pansiot and Rebour 1961: 125); and
2) remove unproductive, dead, or badly placed branches in order that those remaining have good access to light and air (Theophr.

CP 3.2.3; 3.7.6, 9; Pansiot and Rebour 1961: 132 ff.) This facilitates photosynthesis and the intake of carbon from atmospheric CO_2. It also increases ventilation and thus decreases the risk of fungal diseases, and infestations of other pests.

How much pruning is needed depends on the age of the tree. Very young trees need some pruning to shape and structure the tree, but overpruning slows development (Theophr. *CP* 3.7.8; Pansiot and Rebour 1961: 121, 139). Old trees need the most pruning, to rejuvenate them (Pansiot and Rebour 1961: 152–5; Theophr. *HP* 2.7.2, 3; *CP* 2.12.6; 3.7.1). In particular, trees in dry soils, such as those of much of Mediterranean Greece, suffer in productivity if they are allowed to become too large, and will then need more severe pruning (Pansiot and Rebour 1961: 122). This may explain also Theophrastos' (*CP* 3.7.11, *HP* 2.7.2, 3) repeated citation of Androtion that olives require more pruning relative to other trees. Otherwise, productive mature olives which have been well-tended need little pruning, except for the removal of dead, unproductive, or badly placed branches (Pansiot and Rebour 1961: 120–1; Theophr. *CP* 3.7.6, 7). Light pruning, then, was carried out regularly, usually at the time of the olive harvest, when the fruit has been picked. Theophrastos (*CP* 3.7.7) notes that such pruning should be done every other year at most, after the fruit harvest. This occurs during the resting stage of the tree and is also a convenient time when workers are on site and can have a good look at what pruning is required. Heavy pruning was done only as needed. Theophrastos' (*CP* 3.7.4, 7) recommendations on the shape to which trees should be trained and pruned (open in the centre and spreading) and the height allowable (relatively low) are completely in line with the recommendations of modern olive growing (Pansiot and Rebour 1961: 132–3). Interestingly, the reasons given for recommended shaping and pruning are also the same: increased light and ventilation. On the plots of rich landowners, which were largely dug by hand rather than ploughed if they had only trees on them, it did not matter if the lowest branches of trees were too low for draught animals to pass underneath.[13]

[13] Cf. Theophr. *CP* 3.20.1, 7–8. Although land used for arable crops was sometimes dug by hand, generally, ploughing was characteristic of *psilê georgia*, while digging was considered to be normal for land with trees. Modern plots worked by

If pruning and harvesting are carried out together, this entails a saving in labour. Especially when trees are irregularly spaced on scattered plots, some of which (in the case of olives in particular) may be very remote, travel time is saved if the two operations can be combined. In the 'on' years of the olive, the autumn digging may have been done at the same time as well.

Although the olive harvest is a busy time for a farmer who has many trees, big landowners in classical antiquity often hired extra labour at this time. Olive picking can be spread out over several months since, although the fruit continues to ripen, it does not spoil or rot rapidly. However, leaving the fruit for too long before harvesting is likely to lead to deterioration in the quality and quantity of fruit recovered and oil produced. The time pressures of the season more generally (see Fig. 5.4) were exacerbated by the fact that this is also the peak period for the sowing of arable crops: cereals and legumes. For farmers with large numbers of vines, it was the season for trenching and pruning vines as well. Hence, it might have been much easier for wealthy proprietors, who could mobilize large amounts of labour by one means or another, to complete all the necessary tasks within the short season available, than for those who were less powerful and wealthy. The alternatives for poorer farmers would have been either to grow different crops as well as to grow them in a different way.

It is normally assumed that olives were harvested by beating the trees, on the evidence of several Attic black-figured vases, one of which is illustrated in Fig. 5.5. Although this is likely to have been one common method of harvesting olives, there is no good reason to assume these vases show olive harvesting, since almonds, dried figs, or any tree crop not damaged by hitting the ground can be harvested this way. It is also frequently asserted that beating the trees causes alternate year fruiting (Varro 1.55; Blumner 1912: 333; Amouretti 1986; White 1970a). Although alternate year fruiting may be exacerbated in extreme cases, when the branches have been very roughly treated during harvesting, beating the trees is not the cause of the

small-scale farmers with trees and arable combined are usually ploughed rather than dug by hand, so the lowest branches must be high enough not to incommode this operation.

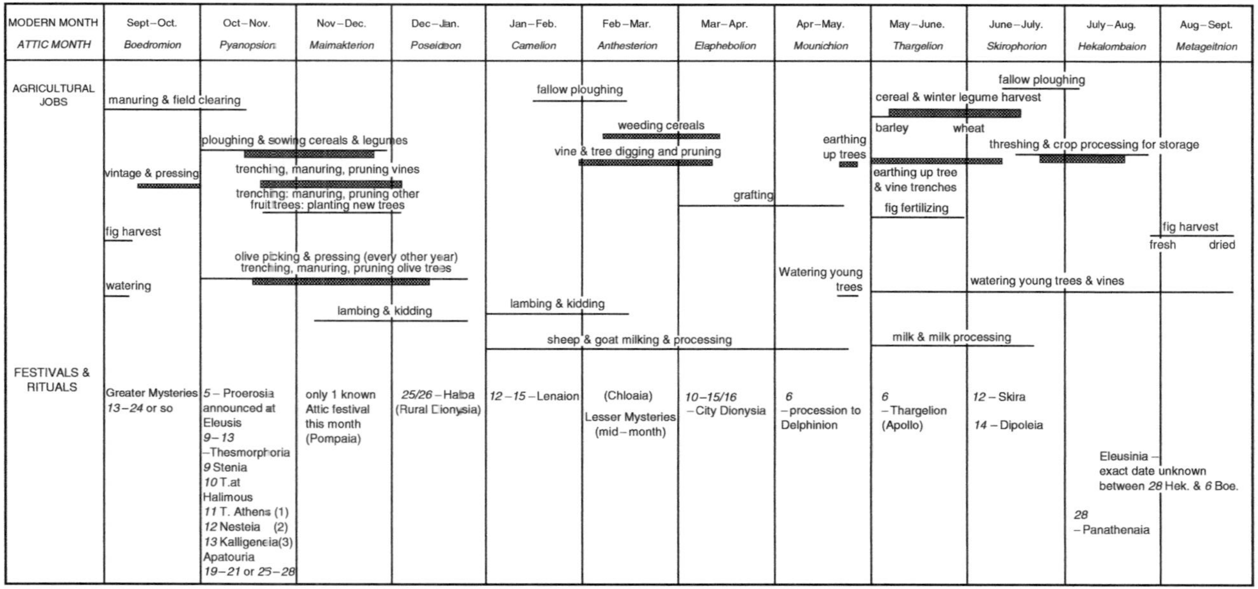

Fig. 5.4. Critical jobs in the ancient Attic agricultural year (L. Foxhall and D. Miles-Williams).

Fig. 5.5. Harvesting a tree crop, Attic neck amphora, sixth century BCE (London, British Museum B226).

biennial fruiting cycle (see Chapter 1.2). Olives can also be picked directly from the trees by climbing them or using ladders. Or, the fruit can be left to ripen and fall of its own accord, and picked up off the ground (this method is most useful for high branches or tall trees). Large cloths can also be placed underneath the trees to expedite the collection of olives on the ground (by picking up the cloth with the olives on it), though this timesaving option was probably restricted to wealthier farmers because of the expense of large pieces of even low-quality cloth. Probably a combination of harvesting techniques was used, and it is likely that these were not limited to the olive, but used for other crops as well.

5.12. CONCLUSION

The techniques of cultivating the olive and other crop trees in classical antiquity are relatively well-documented, even if it is likely that most of our sources provide evidence for the practices of larger-scale and wealthier farmers, who are likely to have had slave labour available. It is extremely difficult to ascertain to what extent the cultivation practices of poorer and smaller-scale farmers resembled those of the wealthy. Cultural beliefs about cultivation practices and techniques were probably widespread across the social spectrum. It is likely that labour availability and the labour requirements of specific techniques are likely to been a crucial factor in how wealthier landowners might have farmed differently from the poor. But choices and decisions which farmers took about cultivation techniques must also have been influenced by a complex range of environmental, economic, and social factors.

6

Processing Olives

6.1. THE PROCESS OF OIL EXTRACTION

6.1.1. Introduction

It is much more difficult to give a complete account of Greek oil pressing than of Roman, and this is reflected in the secondary literature on the subject, which is heavily dependent upon the Roman evidence. Many scholars, beginning with Drachmann (1932) have explicitly or implicitly assumed that Greek techniques were similar to Roman ones, or that they were simply 'less developed' (Ault 1994a; Lohmann 1993: 213–15; Amouretti 1986: 167–75; White 1984: 67–72; Frankel *et al.* 1994: 47–50). Often, scholars use Latin terminology for Greek pressing equipment which is not always exactly equivalent (Hadjisavvas 1992). Even Brun's (2003; 2004a; b; 2005) recent work takes Roman agricultural technology as presented in literary and archaeological sources as the basis of his typologies of pressing equipment (2004a: 5–36) and as the starting point for understanding Greek and other earlier technologies.

Greek farming was generally smaller-scale and less 'professionalized' than the Roman farming best understood from well-documented case studies of large estates in Italy, Egypt, North Africa, and the Levant. Greek pressing equipment is consequently less visible in the archaeological record, and even in the literary and epigraphical record, and it has thus been tempting to retroject Roman approaches and technologies backwards into the Greek world. This approach can lead to serious misunderstanding of Greek agricultural technology and its roles in the economies and societies of the Archaic and

Classical Greek world. The smaller scale of Greek farming overall combined with the rather different role of specialized production in the economies of the Greek world meant that the equipment used was more likely to consist of fairly basic, multipurpose machinery, constructed of multi-purpose components, which might be dismembered when not in use. This is altogether different from the specialized, large-scale permanent installations for processing wine or olives regularly found in the Roman world. The situation is complicated, however, by the fact that pressing equipment is not always easily datable (particularly that recovered in archaeological survey) and that such equipment may have a long, though not necessarily continuous, complex use life.

Contrary to popular belief among scholars (e.g. Cahill 2001: 238), specialized, permanent equipment is not essential for processing relatively small amounts of olive oil or wine, although oil production at any scale above very small amounts without a press would be very slow and wasteful. For wine production, a treading floor is necessary though a press is not, although a press is important for production on any substantial scale. Moreover, while olive and wine processing installations in the world of Archaic and Classical Greece are not unchanging, neither do they 'develop' over time from simple to more specialized and complex technologies in a straightforward way. Even in periods when 'more advanced' technologies were available, simpler ones continued in use. There is considerable regional and chronological variation. This strongly suggests that whether or not equipment which increased the efficiency and speed of processing or saved labour was adopted was closely entwined with specific economic and cultural choices, with local systems of land tenure and agronomic regimes, and with the place of the olive (and the vine) within them. The history of pressing technology is not a story of straightforward evolutionary 'progress'.

There have been several excellent analyses of Roman oil pressing and the machinery involved, most notably the work of Brun (2003, 2004a; b; 2005) and Mattingly (1988a; 1988b; 1988c). Much of the contents of the volume edited by Amouretti and Brun (1993) is devoted to olive processing equipment (mostly from the Roman period or later) from around the Mediterranean. In addition, useful studies of oil extraction in Palestine/Israel (Frankel *et al.* 1994; 1999) based on both archaeological remains and the ethnographic record provide useful comparanda. Much of this work is not directly applicable to Classical, or even Hellenistic, Greece; however, it is very important for understanding the 'cultural biographies' of pressing equipment in the landscape, and the changing place, specialization and scale of olive cultivation over time.

6.1.2. Processing Olives for Oil and the Identification of Olive Presses

Identifying presses that were actually used for olive processing is not always straightforward (Brun 1993). Presses were potentially useful for many operations, both agricultural and 'industrial', and it is probable that many presses were multi-purpose (Brun 2004a: 11, 35–6). I have elsewhere suggested that in order to identify a pressing establishment as being *specifically* for olive processing, it is necessary that not only the press itself be identified, but also the equipment for crushing the olives and separating the oil from the juice and the hot water used in processing (Forbes and Foxhall 1978; see also Frankel *et al.* 1994: 13 for a useful chart of olive processing and Brun 2004a: 5–36). In order to extract oil from olives, they must first be squashed and ground, usually to a pulp. In antiquity this operation was usually carried out in some sort of 'mill' or other crushing apparatus made of stone (often volcanic stone of some kind, if it was available). For crushing olives in small amounts a large mortar might be used. The most widely recognised type of olive crusher is the rotary mill consisting of a bowl with (generally) two millstones commonly known as the *trapetum* (see section 6.4 below). However, for some special oil pressing operations in antiquity, the olive fruits were merely bruised in order that the stones might not be broken (see below, section 6.3). Contrary to what the Romans, at least, thought, crushing the stones does not impair the flavour of the oil (White 1984: 71; Tyree and Stefanoudaki 1996). Most ancient crushing devices, including the ubiquitous *trapetum*, would have crushed the stones along with the olives. At the Hellenistic site of Vrasna in Macedonia, finds of carbonized crushed olive stones from the olives crushed by the *trapetum*-type rotary crushers found on the site show conclusively that these devices crushed the stones (Adam-Veleni and Mangafa 1996). Elaborations in olive pressing technology, some of them not strictly necessary, which increased the time and labour taken to produce oil or reduced the output or fruit/oil ratio, were often associated with perceived higher quality. Olives which had only been bruised or lightly crushed would have yielded less oil than fully crushed ones, and the subsequent pressing operation would have been slower and more difficult. Similarly, green olives yield less oil than fully ripe black olives. The very scarcity of so called 'white' oil (called 'green' oil, *oleum viride*, by the Romans), generally made from green olives, the preference even today for the first cold pressing (so-called 'virgin' oil, another culturally-laden term), and the difficulties of manufacturing oil from partially crushed olives, probably explains why these types were valued more highly. Although some of these elements may well have increased quality (e.g. today the flavour of 'virgin' cold pressed oil is certainly widely preferred because it is less acidic), others may not. The key point here is that the elaboration of the manufacturing process itself becomes a cultural elaboration feeding into the symbolic and social value of olive oil (see Chapter 4.1). That such elaborations rarely become associated with table olives, which are usually much easier to make, is also interesting and significant. One of the few, significantly, is the 'fancy' table olives mentioned by Athenaios, gently bruised in a press before pickling (Chapter 4.1). Again, elaboration in the manufacturing process adds to their perceived cultural value.

Once olives had been crushed, the pulp was placed in bags, basketry frails, or even merely between cloths. These were stacked on the press bed, either on their own or inside a basket or frame. In Classical Greece, presses used for both wine and oil were of the type normally called the 'lever and weights' press or 'beam' press (Brun 2004a: 13–16; see Fig. 6.1). There is no evidence that the screw press was known in Classical Greece or earlier (Brun 2000: 285; 2004a: 17; Drachmann 1932; Paton and Myres 1898).

The famous representation of a lever and weights press on a black figure skyphos in Boston (Fig. 6.2) provides a very clear picture of the pressing operation, in this case probably for wine (Foxhall 1993b: 184; Brun 2004a: 90). This vase offers our earliest visual evidence for a pressing installation, but exactly what kind of pressing is depicted

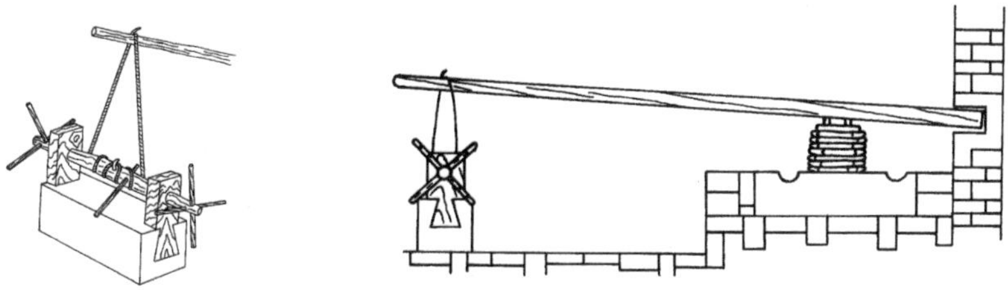

Fig. 6.1. Lever press with capstan (H. Forbes and L. Foxhall).

Fig. 6.2. Lever press, Attic skyphos, sixth century BCE (Boston Museum of Fine Arts 99.525).

here is unclear. Although the picture has often been referred to as an olive press (Amouretti 1986: 167; Isager and Skydsgaard 1992: 64), in fact it is impossible to determine with certainty whether olives or grapes (or indeed, something else) is being pressed. The sacks or frails on the press may well be in a framework of some kind, but it is difficult to be sure of this from the artist's representation. Nor is it possible to see how the press beam is attached—it may well go into a wall. Unusually, this press has its press bed on a stand, with the catchment vessel for the outflow hanging from it. Unless we are not to take the representation literally (and this is perfectly possible), either the stand was stronger than it looks, or the press was not exerting a great deal of pressure, since it would seem that the relatively flimsy stand depicted would not withstand much strain without breaking. This may indicate that the pressing of grape pulp, which needs less force exerted in pressing than olives, is being represented. The identification with wine making is strengthened by the fact that there are many representations on figured vases of aspects of the vintage, most populated by satyrs. The beam is lowered and the pressure applied by means of sacks full of stones tied to the end of it (Amouretti 1986: 167), and extra pressure is exerted by the added weight of a man hanging on to the beam.[1] The later sixth century date of this representation is all the more interesting given that there is only one surviving archaeological example of what is possibly a lever and weights press as early as this from a Greek city, at Klazomenai, discussed below in section 6.3.2. At present this example is unique: the next earliest presses in Greece which are securely datable appear about two centuries later. However, there are similar installations dated to the Iron Age in Palestine (c.1200–568 BCE, Frankel *et al.* 1994: 36–40) and the Late Bronze Age in Cyprus (Hadjisavvas 1992: 3, 21–6) and Crete, and it is likely that the lever and weights press was known in Greece considerably earlier than the sixth century, probably least from the late Bronze Age. The paucity of surviving archaeological examples from the Archaic and Classical periods underpins the probability that Archaic and Classical Greek agricultural machinery was generally modular, ephemeral and relatively small scale.

There are several techniques by which weight can be applied to the press beam in a lever and weights press. The weights themselves are usually made of stone. As in the Boston skyphos, they can simply be suspended from the press beam. Many archaeological examples of weights which are simply large stones with suspension holes have been found, especially in Cyprus (Hadjisavvas 1992). Or, as is more regularly documented archaeologically in Classical Greece, a large

[1] This is reminiscent of the screw press mentioned by Sordinas (1971: 13–4) called the *ai*, because, in order to turn the mechanism to tighten the press, a number of men flung themselves on it shouting '*aiii*'!

Fig. 6.3. Ancient lever and weights press in the mountains, not associated with a farmstead, near a modern vineyard, Methana (L. Foxhall).

rectangular counterweight block can be cut to take the wooden superstructure of a windlass which is then roped to the press beam and used to pull it down (Fig. 6.1). In many parts of the Mediterranean the fixed end of the press beam was held in place by large orthostates.[2] However it seems to have been most common in Classical, and even later in Hellenistic and Roman Greece, to set the press beam in a niche in a wall or carved in a large boulder (see Fig. 6.3).[3] For the lever and weights press, the technique by which a windlass is attached to a press weight, or the fixed end of a press is held in place, may have been more a matter of local/regional preference and tradition than technical efficiency.

The beams of lever presses are usually adjustable. Once the heap had reduced in size, as the oil and juice flowed out, the beam was lowered at the attachment end (see Fig. 6.3, where there are two holes, one below

[2] For North Africa see Mattingly and Hitchner 1993: 446–51; for Cyprus, Hadjisavvas 1992; for Palestine/Israel, Frankel in Frankel *et al.* 1994 42–3: this type was used in northern, but not southern, Israel.

[3] Frankel in Frankel *et al.* 1994 40–1: this type was used in southern, but not northern, Israel.

the other for such an adjustment). Normally, hot water was poured on the olives on the press bed, which released more oil, and the olives were pressed again. This process could be repeated for a third time as well. Beyond a certain point, however, returns diminished and no more oil could be readily extracted from a press of this kind, and the pulp was removed from the press and utilised for fuel or fodder.

The resulting product was a mixture of water, toxic olive juice, and oil. When it was removed from the catchment vessel positioned at the spout of the press bed, it was ladled into large tanks or vats. After it had settled, the oil floated on the surface and could either be skimmed from the top, or, less commonly, the water could be let out from the bottom through a bung or spout. For the amount of liquid that pressing created, the settling tanks had to be very large indeed. Sordinas (1971: 26) found stone settling tanks of 2 m × 1.5 m × 1.5 m (capacity ca. 4.5 m^2) in abandoned eighteenth and nineteenth century oil mills on Kerkyra, and fragments of ones that had obviously been larger originally. Several tanks of this size per mill were necessary for the presses of this period.

6.2. OLIVE PRESSES IN THE GREEK ARCHAEOLOGICAL RECORD

Presses in Classical or later Greek archaeological contexts are generally identified by stone press beds or counterweights (or both). Sometimes floors or platforms coated in hydraulic plaster or walls or boulders with holes positioned to take a press beam also indicate the existence of a press. Few of the Greek presses excavated by archaeologists can be securely identified specifically as olive presses by the presence of crushing and settling equipment nearby. Many are in ambiguous archaeological contexts, in settings which suggest they may have been multi-purpose. Frequently they have been altered and adapted over time. This is particularly evident with excavated presses located in town sites, but can sometimes be detected on rural sites as well.

Pressing installations are found both in towns and in the countryside. Although olive pressing was certainly carried out in the towns, it is likely that much of the processing of both olives and grapes was done in the countryside where the trees were actually growing. In remote locations where there were few roads and paths, at busy times of year like the olive harvest or the vintage, it would put less pressure on human and animal time and energy to press closer to the trees than to haul olives or grapes long distances into a town centre. A good example is the isolated press on Methana illustrated in Fig. 6.3, which was probably largely for grapes, since it is located at a high altitude next to an area good for growing vines at present.

In the following sections I will first examine a selection of the most complete pressing installations located in urban settlements. These are most often the ones which have been excavated, and which therefore have the most secure archaeological and chronological contexts, even if these contexts are sometimes very complex. I have included all of the pressing establishments known to me which definitely date from the fourth century BCE or earlier—these are few in number. However, in the Hellenistic and Roman periods, pressing establishments proliferated within a different political and economic environment than in the days of the self-contained polis. I have examined only a small selection of these, either for comparative purposes, or because in some cases there appear to be earlier components. I will then consider the issues surrounding olive crushing equipment before moving on to examine pressing equipment in Greek countrysides. This has most often been located through archaeological survey, providing less precise archaeological and chronological contexts, but epigraphical sources, notably the Attic Stelae, also provide important evidence. Finally I shall consider the implications for the production and consumption of olive oil in Classical times.

6.3. PRESSING INSTALLATIONS IN TOWNS AND SETTLEMENTS

6.3.1. Introduction

Presses in towns and settlements were almost certainly for the processing of olives close by, within the town itself or near the outskirts. Many may have had other uses in addition to processing olives for oil. Klazomenai provides the only Archaic example. Halieis in the southern Argolid and Olynthos in the Chalchidic peninsula, in northern Greece, are practically the only well-documented Classical examples, both dating to the fourth century. However, the fourth–third century installations at Rachi, a settlement associated with the pan-Hellenic sanctuary at Isthmia and from a large farmhouse on the acropolis of Argilos, probably the centre of a large estate of one of Philip of Macedon's henchmen, are also interesting and important. Considerably more examples date to later in the Hellenistic period, later than the main focus of this work, including Mycenae (Bowkett 1995), Delos, Crete, and elsewhere in Greece (Brun 2004a: 96–111). Of these, only the presses from Delos will be explored in detail, since here it is possible to place them in the context of the Delian countryside to some extent (see Chapter 3.4.2 and below, 6.7).

6.3.2. Klazomenai

Klazomenai (Koparal and İplikçi 2004; Gates 1996: 320–1) has produced the only example of a Greek press from the Archaic period, located in the western sector of the city (Fig. 6.4). It had two distinct phases (Figs. 6.5, 6.6) both falling within the years between the middle of the sixth century BCE and 530 BCE when the Persians destroyed the city, although some of the pits on the site appear to have remained in use up to the fourth century BCE (www.klazomenai.com/isliginikinciev resi_eng.htm). Extensive reconstruction of this press began in 2002 and a working press was operating on the site by November 2004 (www.klazomenai.com/haberler_eng.htm), although there is as yet no final publication of the archaeological remains.

The installation consists of a room or courtyard with fifteen pits cut into the bedrock, not all of which were simultaneously in use. It appears to be quite specialized and relatively complete, with facilities for crushing, pressing and separating olive oil, but could conceivably have been used for processing grapes and other commodities as well as olives. The site is claimed to have the earliest example of a rotary crusher, set in hole 1 (Fig. 6.7; Gates 1996: 320; www.klazomenai.com/isliginikincievresi_eng.htm). However, there are serious problems with the reconstruction of this device (see below section 6.4.1),

Fig. 6.4. View of the pressing establishment at Klazomenai, sixth century BCE (www.klazomenai.com).

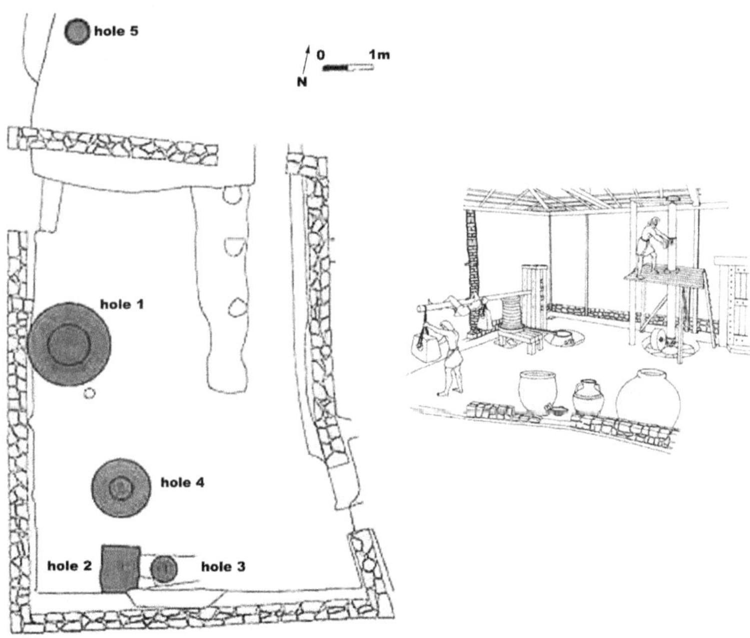

Fig. 6.5. Pressing establishment at Klazomenai, sixth century BCE, plan and reconstruction of earlier phase (after www.klazomenai.com).

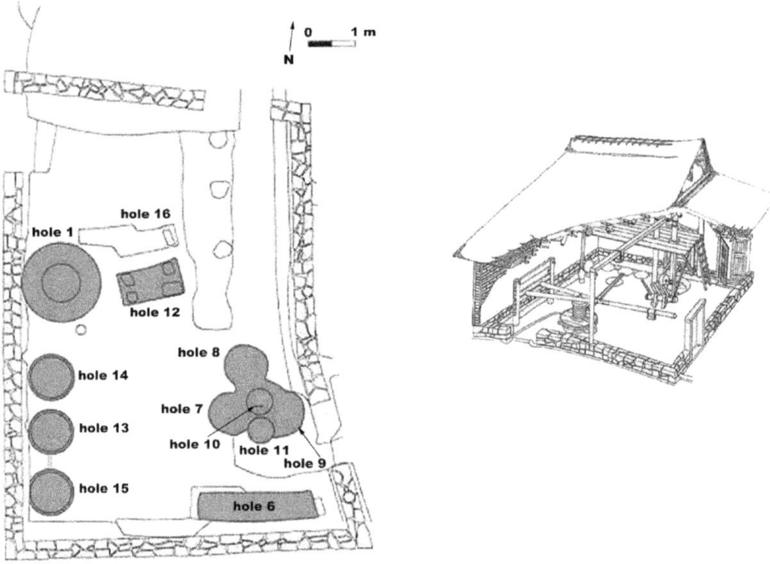

Fig. 6.6. Pressing establishment at Klazomenai, sixth century BCE, plan and reconstruction of later phase (after www.klazomenai.com).

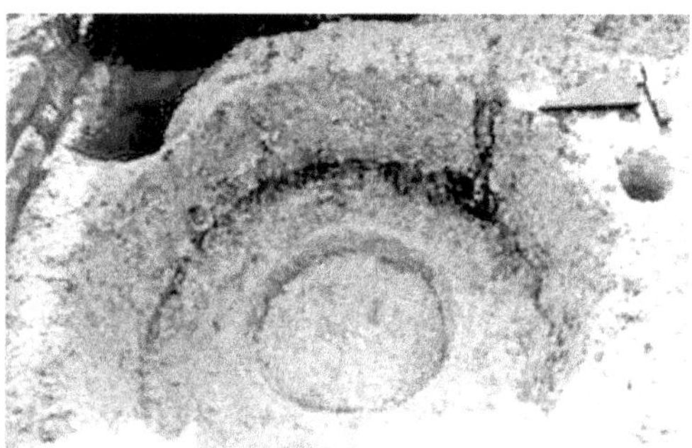

Fig. 6.7. Pressing establishment at Klazomenai, sixth century BCE, Hole 1, supposed seating for rotary crusher. (www.klazomenai.com).

and no millstone belonging to it survives. It seems much more likely that the olives were crushed using a 'roller and bed' type crusher, incorporating the large cylindrical stone roller found on the site (visible in Fig. 6.4, sitting in hole 1), with the shallow, rock-cut trough to the east of hole 1 serving as the crushing platform. Hole

1, which appears to be nearly 2 m in diameter and thus much larger than other known rotary crushers, might therefore have served as a settling tank. There is no surviving press bed. The excavators have restored the first phase of the installation as a small press on a wooden stand similar to the one depicted on the Boston skyphos discussed above, set on wooden uprights fixed in hole 3, with the receptacle set in hole 4 (Fig. 6.5). In the second phase (Fig. 6.6), holes 2, 3, and 4 were filled in and three new settling tanks (holes 13, 14, 15) were created. The excavators have reconstructed the second phase as a lever press operated by a wooden capstan set in hole 12: grooves where the capstan was placed are said to be well preserved (www.klazomenai.com/isliginikincievresi_eng.htm). However, the position seems awkward for the proposed restoration of the press beam. Moreover, the photographs and drawings of the reconstruction appear to have utilized the large stone roller found on the site as a counterweight stone, which was then connected the wooden capstan with ropes. The excavators envisage a press bed set on logs in holes 7, 8, and 9, with holes 10 and 11, said to be cauldron shaped, serving as receptacles. The presses in both phases of the installation, however they should be restored, clearly had numerous wooden parts, including perhaps a wooden press bed. Compared to presses with stone uprights and press beds, the amount of force that could have been exerted was probably limited, and pressing a 'batch' of crushed olives would probably have taken a considerable amount of time (see section 6.8.2). Such relatively slow presses probably did not need a large rotary crusher to supply them with pulp for pressing—the bottleneck would have been at the pressing stage, not the crushing stage. This is a fascinating and important installation, earlier than other, similar ones from the Greek world. Full scientific publication should add much to our understanding of its operation.

6.3.3. Halieis

In fourth-century Halieis several houses located in the Industrial Terrace and the Lower Town have presses in them (Jameson 1969; 2001; Rudolph and Boyd 1978; Ault 1994b; 1999; see Figs. 6.8, 6.9,

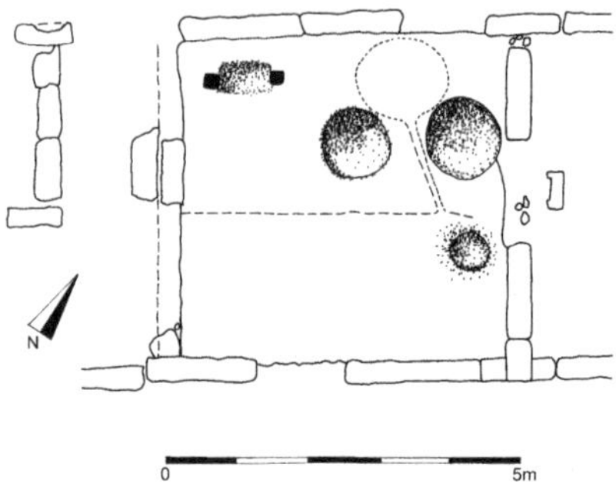

Fig. 6.8. House in Industrial Terrace with press, Halieis (after Jameson 1969: 322).

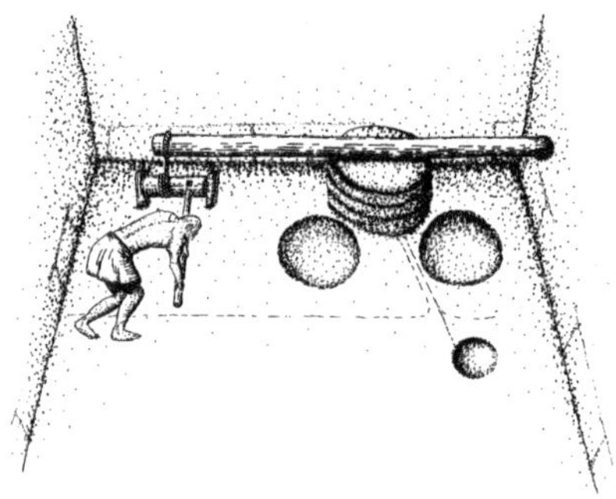

Fig. 6.9. House in Industrial Terrace with press, Halieis, reconstruction (after Jameson 1969: pl. 81).

6.10). It remains the case that only the published plans in the preliminary reports are available, but these installations have recently been discussed by Ault (1999). No rotary olive crushers have been discovered at Halieis in association with or contemporary with the presses (Brun 2004a: 100).

The press in the Industrial Terrace which Jameson (1969: 324; 2001: 283–4) and Ault (1994a: 200; 1999: 652–4) have suggested was for oil is peculiar in many ways (Figs. 6.8, 6.9). This installation consists of a cement-covered area on two levels, the higher one (about 2.50 m wide) 0.03–0.06 m above the lower one (about 2 m wide). On the higher (northwestern) side there is a press bed-like, circular, shallow depression (ca. 1 m in diameter, if the scale of the

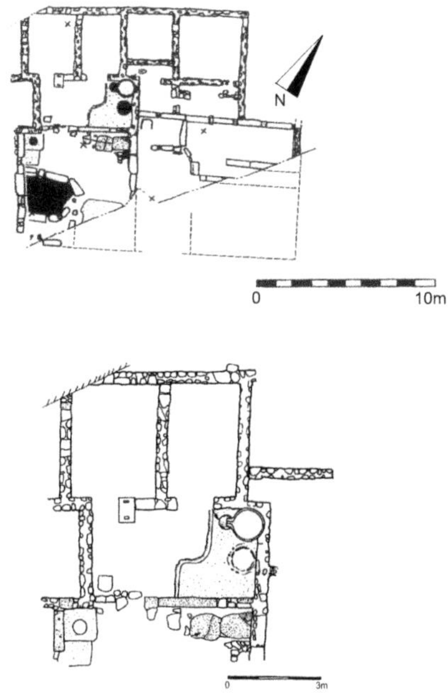

Fig. 6.10. Press in House D, Halieis (after Ault 1999: 561).

published plan is accurate) with a narrow channel leading from the 'spout' to a sunken pithos with its rim at surface level on the lower (southeastern) level. On the higher side, on either side of the narrow channel leading from the 'press-bed' there are two sunken pithoi, which had their rims substantially above the surface of the floor. None of the capacities of these pithoi are given. In the southwestern corner of the higher level, there is a small rectangular depression (very roughly 0.7 m by 0.3 m, as scaled off the published plan), with a rectangular post hole at either end. The post holes are cut through the cement into a block of limestone under the floor, although the stone itself is covered with cement. The size of them is not given.

Jameson (1969: 323) originally envisaged a press fixed between uprights set into the holes in the cement either side of the weight stone. As Jameson (1969: 324) originally reconstructed the press, the beam, running NW–SE between the uprights, would have blocked the doorway in the southern wall leading to a lower, L-shaped room.

In 1993 (Foxhall 1993b: 185–6) I suggested that this arrangement would not work and that the weight stone was attached to a capstan. This interpretation has been accepted by Ault (1999: 563) and Jameson (2001).

Given the small size of the room, any press beam used will have been short. However, if the post holes and depression are reconstructed as a winch mechanism, it is clear that this could be used to pull down the free end of the press beam with a rope, while the fixed end of the press beam attached into the NE wall.[4] There appears to be a gap in the N corner of the room which would take a press beam nicely. This reconstruction would take the beam directly over the press bed-like depression with channel, which makes best sense of the drainage arrangements. However, the paved-over weight stone is unusual, and may suggest that considerable amounts of liquid swilled around the paved area at times. The round depression may well have been designed to take a wooden or stone press bed no longer preserved. The arrangement would allow a short press beam of 3.5 to 4 m in length to be fitted. The pithoi with projecting rims could have been used to hold the substance being pressed, either before or after processing. Perhaps one held the pulp waiting to be pressed, while the other held the residue which had just been removed.

The main difficulty with the identification of this establishment as a dedicated olive press is the lack of crushing equipment and settling tanks in its immediate vicinity. Rollers which could have been used for crushing olives were found elsewhere on the Industrial Terrace (Jameson 2001: 281–2). Such equipment is, of course, inclined to be moved off site in later periods (see below, section 6.5). However, Jameson (1969: 324; 2001: 284) himself originally pointed out the resemblance of this set-up to the installation at Rachi in Isthmia (see below section 6.3.5), and notes that the region was famous for murex production in antiquity. In addition, the substantial plastered area (including the plastered-over weight stone, might suggest that this area was at times used a treading floor for grapes, as well as for a press. I would suggest, therefore, that we need a more flexible

[4] Foxhall 1993b: 186, Ault 1994: 200. Cf. the reconstructed capstan in the Naples Museum, White 1984: 68, fig. 58, or Mattingly's (1988a: 189, fig. 5) reconstruction of a Roman Tripolitanian oil press, but turned 90 degrees.

interpretation of this installation as multifunctional. It could have been adapted over the short term, or more probably seasonally, for the processing of both oil and wine, and it may have served other industrial functions as well—dyeing is one possibility, though probably not for murex—the lack of murex shells almost certainly precludes this use (Ault 1999: 560). However, the lack of archaeologically detectable crushing equipment suggests that if this press was used for olives, oil processing was on a very small scale.

The same issue of crushing facilities arises with the presses in the Lower Town (Jameson 1969: 328; Rudolph and Boyd 1978: 344–52; Ault 1994b; 1999). There is a press in House D, and possible presses and/or grape treading floors in House A, House C, and the house in area 4. For most of these, the lack of settling and crushing facilities suggests these are not designed *primarily* for processing olives, as the excavators suggest (Jameson 1969: 328; Rudolph and Boyd 1978: 350, who only noted the press in House D), although they could have been used for processing both olives and grapes as well as for other 'industrial' uses.

The press in House D (Fig. 6.10) is the best preserved of these, but the entire house was not excavated, so the full range of facilities which accompanied the press may not be exposed (Boyd and Rudolph 1978: 350; Ault 1994b: 198–200 and Fig. 2; Ault 1999: 560–2 and figs. 4 and 11). The southernmost area of the house, probably part of a courtyard, has a large, irregularly shaped receptacle, lined with stone slabs but with an earthen floor (capacity 5 m^3), identified by Ault (1994b; 1999) as a 'kopron' (see section 7.7 for an alternative interpretation of these features). There is a well with a limestone wellhead to the NW of this vat, and immediately to the E of the well, in the N corner of the courtyard, is a stone trough, 1.0 × 2.30 m in size, and set 0.10–0.20 m deep into the floor. This is identified by Ault (1999: 562) as a 'crushing trough', although only relatively small quantities of olives could be processed at one time. The room opening off this room to the N contains a good limestone or marble press bed with sunken basin (depth 0.28 m; diameter 0.42 m) and at right angles to it a larger sunken pithos (depth 0.78 m; diameter 0.70 m) on the NE side set in a cement floor. The press bed looks as it could be turned to expel liquid into either receptacle. The rectangular weight block to operate the lever press is located 4 m to the W of this installation. In its present location it cannot be *in situ* for operating the press, as it is partly behind a wall, which would make connecting the press beam to the winch mechanism impossible. This suggests that either the block was moved after the house was abandoned, or that the press went out of use while the house was still occupied.

It is difficult to identify this as a dedicated olive press with certainty. The 'crushing trough' identified by Ault (1994b: 199–200 and n. 14; 1999: 562) is not entirely satisfactory. Ault erroneously supposed that the crushing process removed the stones and does not explain how he thinks it was used for crushing olives. The recessed bottom might be a problem for using it as a 'roller and bed' type crusher here, and no roller was found, although it could have been used as a crushing surface for pounding olives by hand, as in a mortar. Although this installation may well have been used for pressing olives, it is not at all clear that was its only function, or that it operated as an olive press over its entire lifetime. Given that it seems to be part of a fairly large and elaborate establishment, it seems more likely that other 'industrial' operations were carried out in addition to grape or olive pressing.

6.3.4. Olynthos

The planned townscape of the North Hill of Olynthos was built in the later fifth century BCE (Cahill 2001: 38–40). The city was destroyed by Philip of Macedon in 348, but numismatic and other archaeological evidence suggests that some neighbourhoods, including some of the houses with pressing equipment, were reoccupied after the sack of the city down to 316 BCE (Cahill 2001: 49–61). House A xi 10 at Olynthos (Fig. 6.11) includes an installation identified by the excavators as an 'olive or grape' press (Robinson and Graham 1938: 339, pl 82, 102; 129). The identification of this facility as an olive press is upheld by Cahill (2001: 239–41). However, this identification is not straightforward although the installation appears to be relatively well preserved. It is located on the south side of room j, which has a cement floor with two square holes against the wall left apparently to take the uprights to support the fixed end of the beam. But on closer

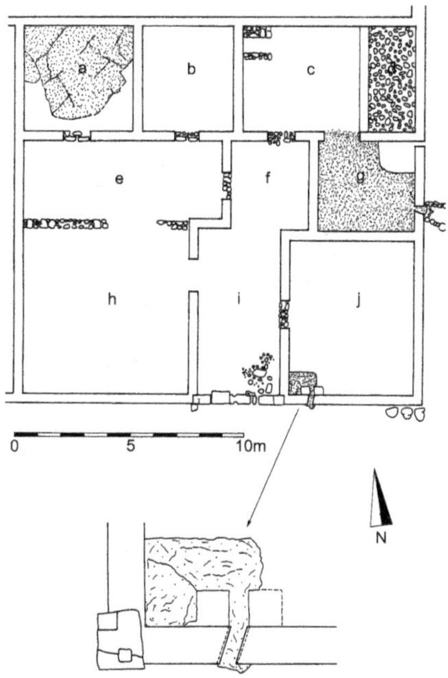

Fig. 6.11. Alleged 'olive or grape press', House A xi 10, Olynthos (after Robinson and Graham 1938: pl 82.1, pl. 102 and Cahill 2001: 241, fig. 50).

inspection, this is rather an odd press. The pithoi embedded in the floor of the room are not set up to take the runoff from the press, and there is no other feasible collecting basin, nor even a place for one. Nor is there any sign of a press bed, nor even a place where a press bed might have been, for the cement paving does not seem to extend far enough to the north to take one here, even had the pressbed been made of wood, and the uprights are very close together (they appear to be about 20 cm apart in Cahill's [2001: 241, fig. 50] plan). Finally, the drain between the two holes which are supposed to take the press uprights is very peculiar—one false move and all the oil (or whatever) lands in the street.

That this house contains some kind of 'industrial' establishment I think is certain, but it is hard to guess what kind of work was carried out (except that it was probably wet). The sizeable courtyard, the entrance designed to take wheeled vehicles, and the cement and pebbled surface of room g, which adjoins the 'press' room, j, all point in this direction. It is possible that the facility was used for several different activities, and wine production could conceivably have been one of them. Brun (2004a: 100) suggests that the paving and holes in room j were designed to take a portable wine press of the type represented on the Boston skyphos (see above, section 6.1.2), though there is no obvious treading floor preserved. Small-scale dyeing or fulling is another possibility suggested by the substantial number (57) of stored loomweights found in the house (Cahill 2001: 174, 176, 177–8, 241): in addition to their normal function, loomweights can be used to keep wet, dyed cloth or yarn stretched out to prevent shrinkage after dyeing or other treatment. However, the peculiar size and layout of the 'press', the lack of a counterweight stone for a press, and the lack of crushing and settling facilities, all suggest that it was not a dedicated olive press, at least in its preserved final state. Even the operation of the 'press' itself remains somewhat mysterious—it certainly does not seem suitable for manufacturing oil in any quantity.

Another possibility is that the preserved facility was a later adaptation of an earlier installation, which might have been a more typical press. House A xi 10 appears to have been one of those occupied down to 316 BCE judging from the numismatic finds (Cahill 2001: 54, 59, 60–1). It is possible that the reoccupiers of the site adapted an existing installation to some other purpose, and in the process got rid of redundant equipment which might have revealed its original function. It is possible that any useful tanks and crushing equipment belonging to an earlier phase might have been recycled elsewhere. However, a counterweight block, had it existed, is more likely to have remained in the house even if no longer used because its sheer size and weight would have made transportation difficult.

In 1993 I identified the facilities in House A 6 (Fig. 6.12) as a possible olive press (Foxhall 1993b: 190–1), and this identification has been broadly accepted by Cahill (2001: 241–4). Like House A xi 10, this house had a large court with double doors suitable for wheeled vehicles (Robinson 1930: 68–74; Robinson and Graham 1938: 75–6; Cahill 2001: 241–2), suggesting it served as a base for some sort of commercial or 'industrial' establishment. The house itself had expanded to take in the northern half of House A 7 and the original gap between the two houses, to the south (Cahill 2001: 242).

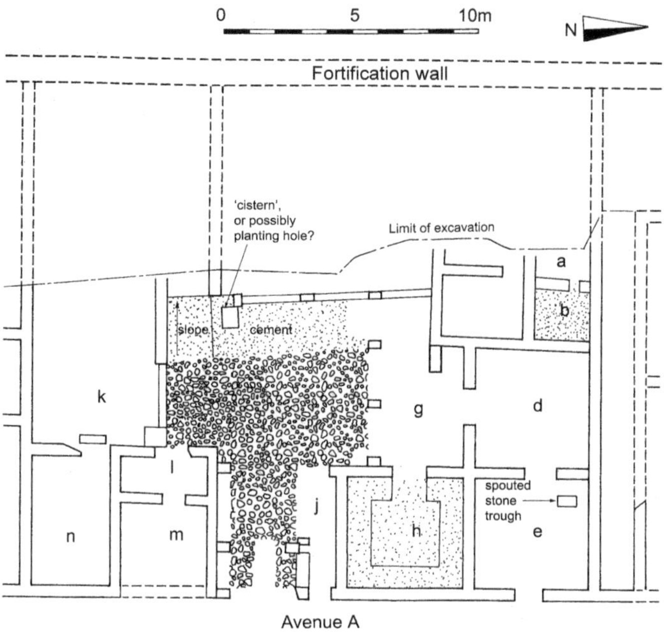

Fig. 6.12. House A6, Olynthos (after Cahill 2001: 242, fig. 51).

Previously I suggested that the small, cement-floored room b, identified by the excavator as a bathroom, could be a press room (Foxhall 1993b: 190). In the excavator's relatively clear photograph (Robinson 1930: fig. 192) a circular space in the cement flooring, just in front (west) of the buried basin is visible. This looks as if it could have taken a press bed. On the northwest side of the room is a large rectangular block with a groove around it, which the excavator reconstructed as part of the western wall of the room. This could be a counterweight for a press, but if so, it is the only one identified on the site. In addition, the western section of the courtyard is covered in cement, and a single millstone from an early type of rotary olive crusher (see section 6.4) was also found in the courtyard. Cahill (2001: 243–4) prefers to keep the identification of room b as a bathroom, since it is typical of the 'bathrooms' regularly found as part of the 'kitchen complexes' elsewhere on the site, with depressions to hold a bath tub. He has instead suggested that a press was located on the cemented strip on the west side of the courtyard, and that a collecting basin might have been robbed out on the north end (Cahill 2001: 242–3). The difficulty with this idea is that there is no evidence in this area of a press bed (or an area where a press bed might have been) or a counterweight for a press, nor is there any archaeological evidence of how a press might have been fixed to the floor or a wall in this area (e.g. a hole in a wall to take a press beam, uprights to support a press, etc.). Moreover, although the millstone from a rotary olive crusher is present, the bowl of the crusher is absent. In summary, it is possible that there was once a functioning press in this house, perhaps with substantial wooden elements which have not survived, but the evidence is inconclusive.

In addition, the courtyard contained seven hopper-rubber mills and the lower stones of five saddle querns, which Cahill (2001: 244) assumes were for grinding grain. This could be correct, but grain mills of both the types represented here were used for grinding many other things as well, and it is puzzling that there are no upper stones for the saddle querns. If all of these millstones are be associated with the processing operations carried out in this house, it is not at all clear what they were. As in the case of House A xi 10, it is possible that the visible archaeology represents the adaptation of earlier facilities for new purposes. It is also possible that the facilities were used for more than one operation.

Cahill (2001: 244–6) has proposed another possible pressing establishment in the adjoining Houses A vi 8 and A vi 10 (Fig. 6.13). These houses appear to have been among those reoccupied after the 348 BCE destruction of Olynthos. In the final phase of these houses, it seems that House A vi 10 had expanded to the west, and had taken over the SE corner of House A vi 8 for a dining room. The rest of House A vi 8 had few finds of domestic equipment, and may at this stage have served as a workshop and storage area for the inhabitants of House A vi 10 (Cahill 2001: 244–5). Room g of House A vi 8 has a roughly square cemented area in its SW corner, emptying into a receptacle holding about 110 L. This is likely to be a wine treading floor (Brun 2004a: 100). There is a smaller area of cement flooring to the east of this—perhaps for a facility which had gone out of use, since it appears to have no obvious function in the existing set up. In house A vi 10, room k contained a large rectangular cement platform

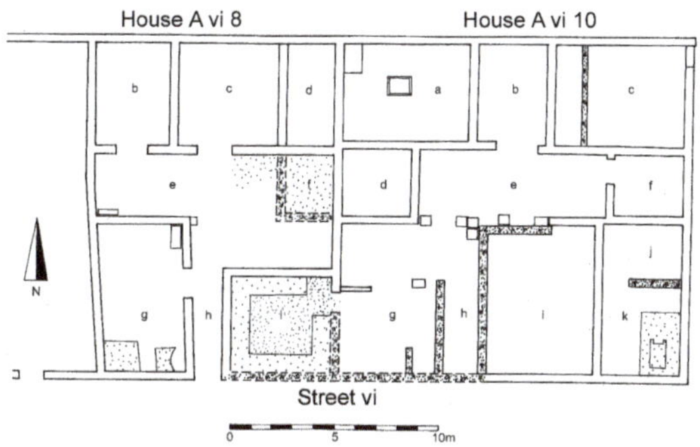

Fig. 6.13. Houses A vi 8 and A vi 10 (after Cahill 2001: 245, fig. 53).

with a rectangular hole in it. Cahill (2001: 245) suggests that the platform is part of a press, and that the rectangular hole is where a press bed was removed. However, this rectangular hole is not shaped like a press bed, and the press beds from Olynthos, such as the one found in House A 1 are all circular, not rectangular (Robinson and Graham 1938: pl 81, 82). The stone mortar buried in the floor of room i could have been used for olive-crushing as Cahill (2001: 245) suggests, but only for relatively small amounts of fruit; of course such mortars could, and probably would, have served many functions. Neither house has features such as settling basins, large pithoi, settling tanks, or counterweight stones which one would expect to find with a working press. As in the case of the other Olynthos houses with processing facilities, the evidence for a olive press is weak, although it is likely that wine making and other processes were carried out in these buildings. As in the case of House A xi 10, reoccupation may have clouded the archaeological record of agricultural processing and 'industrial' activities.

6.3.5. Isthmia, the Rachi Settlement

The Rachi settlement sits on a long narrow ridge to the south of and above the Temple of Poseidon at Isthmia, and just southwest of the Later Stadium. It was occupied from the second half of the fourth through the end of the third centuries BCE. The settlement comprised

Fig. 6.14. The Rachi settlement, Isthmia, site plan (Anderson-Stojanovič 1996: 64, fig. 4; courtesy of the Trustees of the American School of Classical Studies at Athens).

a cluster of about eighteen houses which included workshops and a large structure (North Building) with five big rooms, to the north of a narrow street separating it from the main group of excavated houses (Fig. 6.14). It is clear that agricultural and 'industrial' processing activities were carried out in many parts of the site (Anderson-Stojanovič 1996).

Anderson-Stojanovič (1996: 66–7, 91–2) has argued that the 'workshop areas' of most houses were press installations for manufacturing olive oil, and that six identifiable presses survive (Anderson-Stojanovič, forthcoming). Many features were cut into the bedrock. There are no stone press beds from the site, but circles of pebbles set in the cemented platforms in houses XVI and XVII (Anderson-Stojanovič, forthcoming), may indicate that wooden press beds were used. Houses III and IV, recovered in Broneer's (1955: 124–8; 1958:

17–20, 31, 32, 36) excavations in the 1950s, contain the two best preserved presses, but detailed plans of these have not yet been published.

The installation in House III, room A (Fig. 6.15) consists of a rectangular cemented platform, roughly 1.75 × 1.50 m in size (as derived from the plan in Anderson-Stojanovič 1996: 64, fig. 4) with two receptacles on the northern side. The platform drains into the smaller (western) of the two receptacles. A circular (stone?) object appears on Broneer's (1955: pl 49b) photograph, with a smaller, rectangular stone in front of it. The photo is not clear enough to be sure of the function of these, but it is possible that the circular stone was designed to sit on top of the pile of frails during pressing. Broneer (1955: 126) mentions a large ceramic jar located to the west of the platform, which appears in his photograph, with the

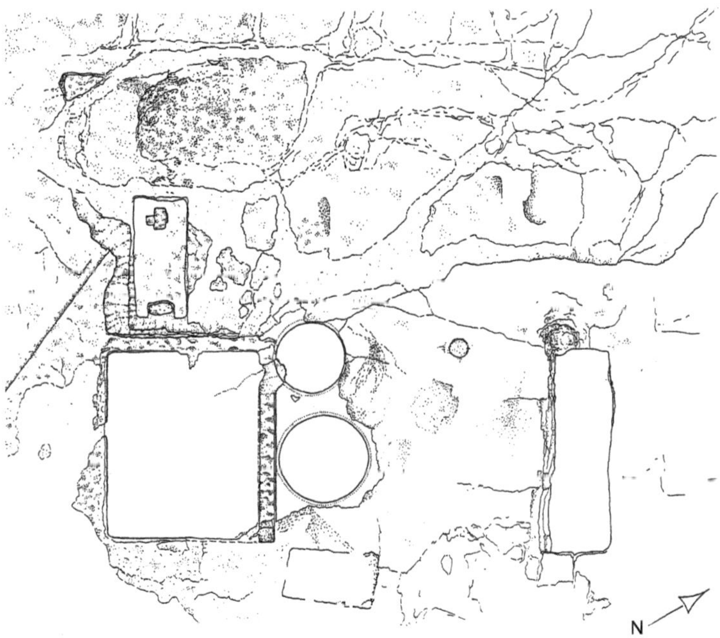

Fig. 6.15. The Rachi settlement, Isthmia, press in House III (V. Anderson-Stojanovič; courtesy of the Trustees of the American School of Classical Studies at Athens).

counterweight block with two mortise holes in the upper surface identified by Anderson-Stojanovič (1996) visible underneath it. According to Anderson-Stojanovič (forthcoming), it is clear that the function of the installation changed over time since the counterweight block was covered with earth before the pithos was placed on top of it. The position of the counterweight block suggests that the press beam was fixed in the southeast wall of the building, and that the beam ran southeast–northwest over the cemented area parallel to the counterweight (Anderson-Stojanovič, forthcoming). A small rounded niche in the southeast wall visible on the plan of house III (Fig. 6.15), directly in line with the counterweight is likely to have taken the fixed end of the press beam. If this interpretation is correct, the press beam was very short, no longer than 1.50–1.75 m. The press would therefore not have been a very powerful one. At the northern end of this room there is a rectangular tank cut into the rock, approximately 1.75 m × 0.40–0.70 m, which would have served as an appropriate settling tank.

It is probable that this facility was used for small-scale olive pressing, even though no obvious crushing equipment is present. However, the cemented platform would have served well as a wine treading floor as well (Brun 2004a: 103), and the press could also of course have been used for wine. It seems most likely that the functions of this installation included winemaking as well as olive oil production in small quantities. However, its function need not have been limited to these activities.

The installation in House IV appears to be similar (Fig. 6.16). Here the large square cemented platform measured about 2.60 m per side (Broneer 1958: 19). Adjoining it to the west was a smaller area of better quality and better preserved plaster flooring measuring 1.80 × 1.90 m (Anderson-Stojanovič, forthcoming). There are two circular, cement-covered basins, the lower sections cut into the bedrock but the upper sections build in masonry. Broneer suggested that the three cuttings in the floor found filled with stones to the north secured a press, but the size and position make this seem an unlikely possibility—it is more likely that they secured a winch or capstan for a press, even though there is no surviving counterweight stone (Anderson-Stojanovič, forthcoming). A small niche visible in the southwestern wall on the plan (Fig. 6.16) might have taken the fixed end of the

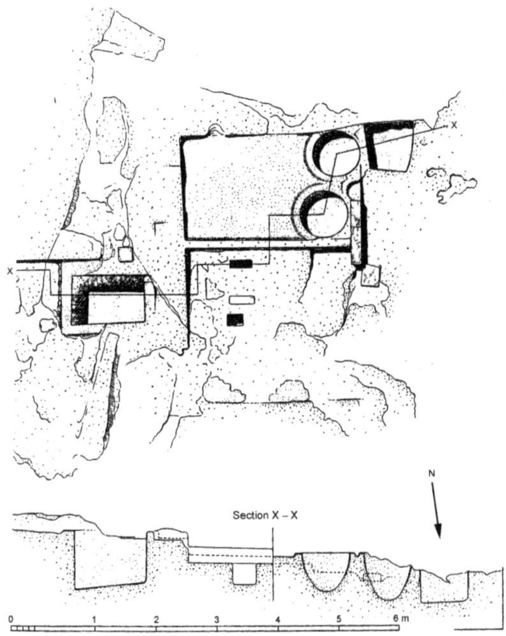

Fig. 6.16. The Rachi settlement, Isthmia, press in House IV (V. Anderson-Stojanovič; courtesy of the Trustees of the American School of Classical Studies at Athens).

press beam. This would allow for a press beam of just under 3 m in length (Anderson-Stojanovič, forthcoming). The installation would have worked well as a treading floor for wine. It could certainly have been used for pressing olives although there is no surviving stone press bed and there are no obvious olive-crushing facilities or settling tanks present. This suggests that if it had been used for pressing olives, it was on a relatively small scale. As in the case of House III the installation could have been used for other processes as well. The stones found filling the holes in the floor and the difference between the cement of the two cemented areas might suggest that the installation, and the uses to which it was put, were altered over the course of its lifetime.

Although there are a number of interesting installations across the Rachi site, it is difficult to be certain that any of these was exclusively

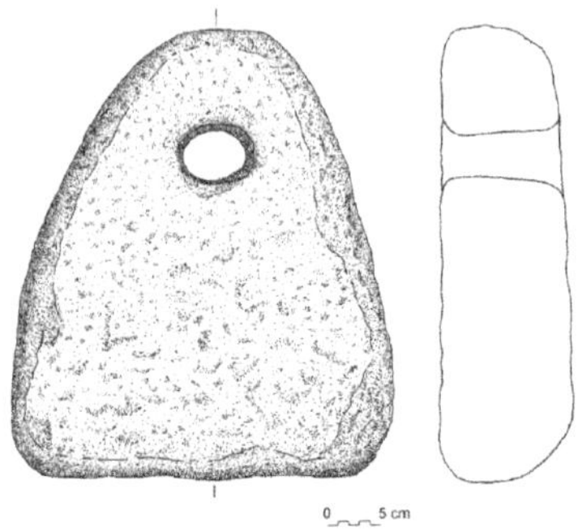

Fig. 6.17. The Rachi settlement, Isthmia, stone press weight (V. Anderson-Stojanovič; courtesy of the Trustees of the American School of Classical Studies at Athens).

used for pressing olives. None of the presses seems to be very large. At an earlier stage in the investigation of the site it had been thought that these facilities were for dyeing (Kardara 1961). Fulling and tanning were other possibilities considered by Anderson-Stojanovič (1996: 91), and later rejected by her (Anderson-Stojanovič, forthcoming). Hanging press weights (Fig. 6.17) were found, although there is little identifiable crushing equipment. Fragments of a circular millstone and the central column from the mortar of a rotary olive crusher are reported as well as two possible rollers from 'roller and bed' type crushers (Anderson-Stojanovič forthcoming). The paucity of obvious olive-crushing equipment directly associated with presses is interesting and suggests that any oil manufacturing was relatively small scale, and that these installations were used for other activities as well. Anderson-Stojanovič (forthcoming) observes that there are many pits and depressions cut into the exposed bedrock of Rachi which might have been used for olive crushing, but if this interpretation is correct, it also emphasizes the relatively small scale of oil production. It is clear from the archaeological remains themselves that the facilities were changed and adapted for new uses over time. It also seems likely, as Anderson-Stojanovič (1996: 92–3) suggests, that whatever production was carried out here, was related to consumers using the sanctuary. Certainly there seem to be too many processing facilities simply for the needs of the small settlement on the Rachi ridge. Olive oil for both athletic use and food consumption is certainly one possibility, but wine and other commodities would also have been desirable and would have found a ready market with visitors to the sanctuary.

6.3.6. Argilos

Argilos is located in Macedonia, near Amphipolis. The city was sacked by Philip of Macedon in 357 BCE, but the acropolis was reoccupied by at least one substantial structure that appears to be a public building, and on the summit by a large and solidly-built farmhouse that may have been the headquarters for a large estate given to one of Philip's '*hetairoi*'. Occupation continued throughout the third century BCE, but the city was then abandoned and was not apparently reoccupied in the Roman or Byzantine periods (http://www.argilos.org, accessed 12/5/2006; Bonias and Perreault 1993; 1997; Touchais *et al.* 2000).

The Hellenistic farmhouse (Fig 6.18) contained a well-preserved rotary olive crusher (*trapetum* type) (Bonias and Perreault 1997: 544–8), the mortar (external D. 1.54; H. 0.53 m) of which appears to be seated on a platform in the central room on the eastern side of the house. The central column of the mortar is fully preserved (D. 0.70 m) with a circular hole in the centre 0.15 m deep. The two millstones which belong with the mortar were found in the courtyard. These appear to be smaller in diameter and considerably more rounded than the fourth century millstones from Olynthos and elsewhere (see below, section 6.4.1). A large pithos, 2.5 m in length was also found in the courtyard lying on its side. On the north side of the courtyard is a feature which may be a settling tank. Outside the house a rough press bed was found. However, no cemented area of the sort typical of press installations from other sites, press weights, or holes to take a press beam were recorded (Perreault, pers comm.). At present, the pottery from the site is not published.

The presence of a rotary crusher in the absence of unequivocal evidence for a pressing installation is unusual. This is a puzzling installation in many ways, and were it not for the insistence of the excavators that the site was not occupied in Roman times, it would be

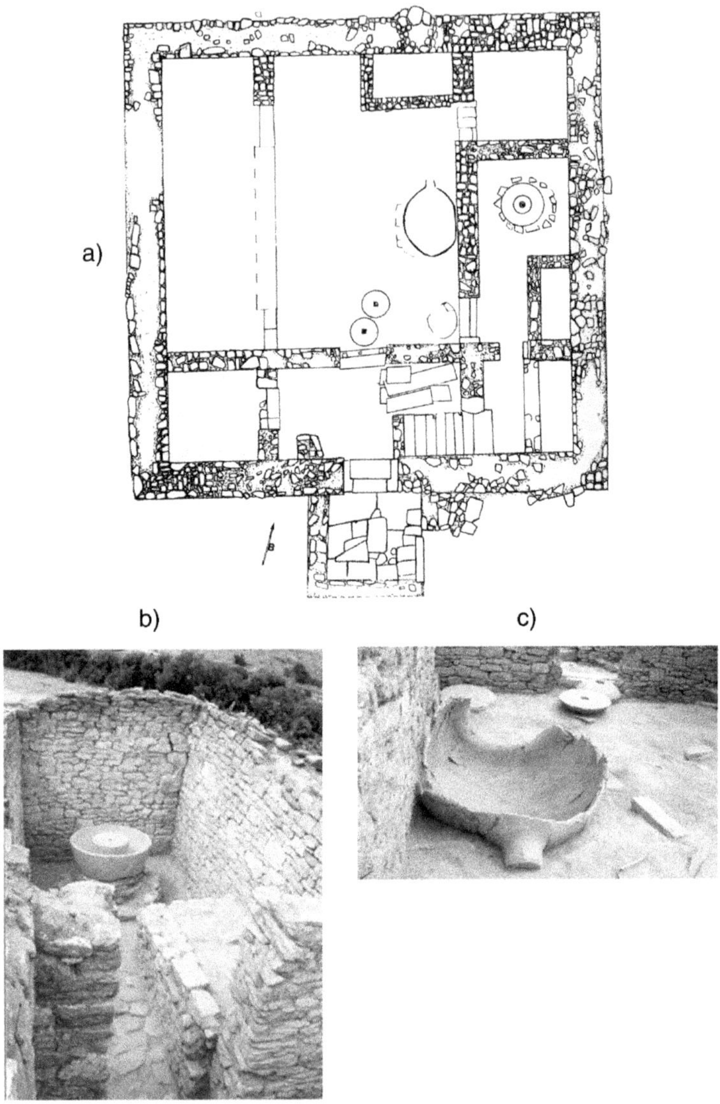

Fig. 6.18. a) Argilos, plan of farm house (Building A), fourth–third century BCE; b) mortar of rotary crusher; c) millstones from rotary crusher and large amphora (Touchais *et al.* 2000: 945, figs. 223, 224, 225, reproduced by permission of the excavators, Z. Bonias and J. Perreault).

tempting to attribute this rotary olive crusher to Roman or Late Roman reoccupation. The closest parallels for the shape of the millstones seem to be the third–second century BCE stones from Maresha in Israel (Kloner and Sagiv 1993: 120–2), although the stones there were used singly, not in pairs. However, if this rotary crusher is genuinely early, it is not surprising that it is found in a structure that must have been used for processing the crops of the large estate of a wealthy Macedonian.

6.3.7. Delos

Presses and pressing equipment from the complex urban site of Delos have been thoroughly studied by French scholars (Brun 2004a:108–13; 2000; 1999; Brun and Brunet 1997; Bruneau and Fraisse 1981; 1984; Blackman 1998: 105–8). It is not surprising, given the extensive development of the city in the Hellenistic and Roman periods, that no pressing equipment or installations of the Classical period have been identified. Three later pressing establishments have been particularly well published: 1) the press in Street 5, 2) the first century BCE presses in House III O in Theatre Quarter; and 3) the presses in House I B in the Stadium Quarter. Although these are all later than the Classical period which is my main focus here, these meticulously investigated presses provide useful comparative data for understanding the survival and adaptation of pressing equipment in urban settings. Moreover, the information provided by the leases of the estates owned by the Temple of Apollo on Delos, Rheneia, and Mykonos, as well as the archaeological investigation of the countryside by French teams (Brunet 1990a; 1990b; 1999; Brunet and Poupet 1997), can be analysed in tandem with the evidence of the urban archaeological remains.

The press in Street 5 (Fig. 6.19) was originally published as a wine treading floor dating to the some time between the middle of the first century BCE and the Imperial Period on the basis of the stratigraphy and the relative chronology of walls in this sector of the city (Bruneau and Fraisse 1981). Unfortunately at that time the authors had not taken into account the large counterweight weight block decorated with Christian symbols (crosses, peacocks and fish) situated in front

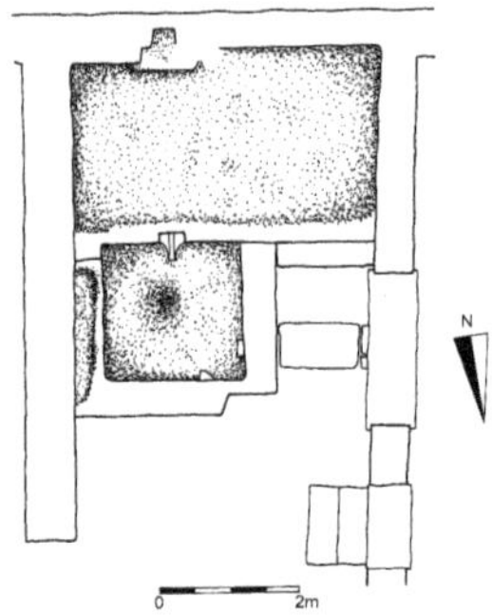

Fig. 6.19. Delos, press in street 5 (after Brunet and Fraisse 1981: 136, fig. 10).

of the installation. Bruneau and Fraisse (1984) then reconsidered the chronology of this installation and put forward several alternative solutions, including the possibility that it was used in more than one period. The upper cemented platform to the north would certainly have served well as a wine treading floor, with the juice emptying into the lower container. No press bed was found on the site, nor was there evidence of olive crushing equipment, although it is possible that a container located in area C could have functioned as a settling tank. This installation seems to be a classic example of a multi-period, multi-purpose facility; perhaps originally built as a treading floor, it may have been adapted in Late Antiquity for use as a press. The press could have processed wine or oil, and the cemented platform could still have been used as a treading floor.[5]

The pressing installation in House III O in Theatre Quarter was most recently investigated by Brun and Brunet (1997; Blackman 1998: 107), who identified it as a dedicated oil press (Fig. 6.20). It was constructed in the first quarter of the first century BCE in a building which in its earliest form dates back to the fifth century BCE. Around the time the press was built, House O in which it was situated was joined to House P. The press had probably gone out of use by the middle of the first century BCE or shortly afterwards. This is an elaborate installation, with two press beds (only the southerly one

[5] Brun (2000: 284) identifies this installation as dating only to the Early Christian period.

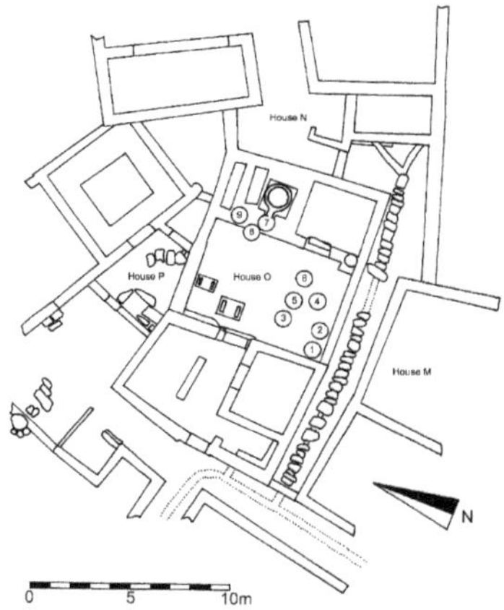

Fig. 6.20. Delos, press in Theatre Quarter, House III O (after Brun and Brunet 1997: 584, fig. 8).

survives) sitting side by side on a paved platform operated by counterweight blocks at the west end of the room. If these are *in situ*, then the northernmost press beam (restored at 7 m) was shorter than the southernmost one (restored at around 8 m). Brun and Brunet (1997: 598) have suggested that the northerly press was used for the first pressing of 'virgin' ('white') oil, while the longer southerly one, which would have exerted more force, was used for producing lower grades of oil. This would also explains why the large receptacles are situated so that the southerly press drains into them, since the olive pulp would have had large quantities of hot water poured over it at this stage of the processing.

Six pithoi are located to the south of the press installations, each of about 650 L capacity, giving a total capacity of 3900 L (Brun and Brunet 1997: 602; Brun 2004a: 111). When the bottoms of these were excavated numerous charcoal samples and a number of whole olive stones were recovered. The charcoal consisted of a mix of brushwood (*Quercus ilex*, *Q. coccifera*, both evergreen oaks, as well as deciduous oak species and buckthorn, *Rhamnus*/*Phyllirea* spp.) and fruit tree prunings (apple and pear, olive and vine) (Terral in Brun and Brunet 1997: 610). The latter are likely to have been collected in the late autumn, when the trees would most likely have been pruned. How the carbonized wood ended up in the bottoms of the pithoi presents an interesting enigma. It is possible that they fell from an upper story or that the room served in part as a wood store in its latest phase.

The analytical report suggests that all of the olive stones found in the bottoms of pithoi 4, 5, 6, and 8 were whole (Brun and Brunet 1997: 596; Terral in Brun and Brunet 1997: 612). If so, this might suggest that at least in the final phase of the building's life that the pithoi were not used for the storage or processing of olive oil—it is of course possible that all the organic remains discovered date to a time after the press had gone out of use. Olive stones from the residue of oil processing or pulp for pressing would almost certainly be fragmentary. Whole stones could indicate that table olives were stored in the jars at this stage.

No crushing facilities are present in this installation. Brun and Brunet (1997: 597–8) suggest that for a press of this size there must have been some kind of dedicated olive crusher, perhaps located in room f to the south of the press beds, but removed at a later date.

This installation appears to have gone out of use, or at least its function might have changed, before the building in which it was situated came to the end of its use/life. It is interesting that such an elaborate pressing establishment operated over such a short period—no more than fifty to seventy five years. Its use for the production of olive oil seems likely, although it could have been used for wine as well.

The installation in House I B (Fig. 6.21) dating to the late second–early first century BCE in the Stadium quarter was discovered in the old excavations early in the twentieth century (Plassart 1916) and reinvestivated by Brun in 1997 (Brun 1999; 2000: Blackman 1998: 105–7). Brun has identified this installation as a perfumery equipped with four 'furnaces' and a pair of wedge-presses which he has restored on the basis of much later first century CE Roman iconographic evidence from Pompeii and elsewhere. Two elaborate but relatively small press beds were found (pressing surface 0.63 m), but both had been reused upside down as floor slabs in a later phase of the building after the presses had gone out of use, before the middle of the first century BCE. The furnaces were maintained in this later phase, perhaps as part of a bath. Brun has restored the position of the pressbeds as set between two pairs of holes for uprights in a paved area at the east end. There are no traces of counterweight blocks, collecting receptacles, storage vessels, mortars, or crushing facilities.

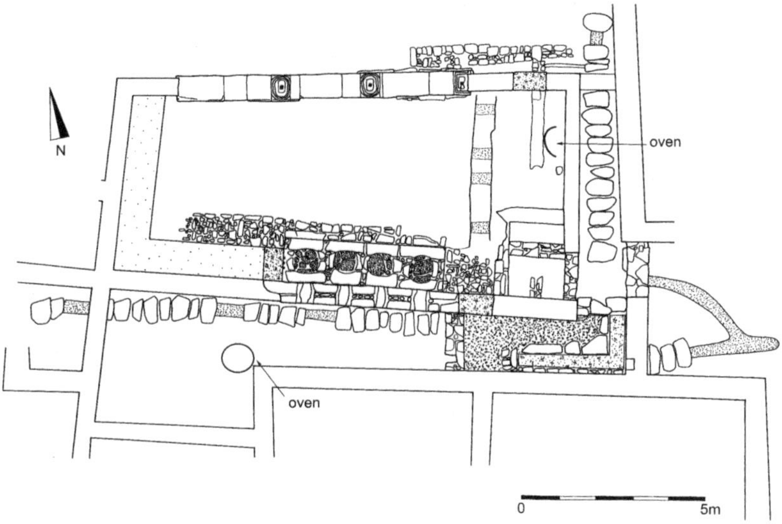

Fig. 6.21. Delos, press in Stadium Quarter, House I B (Brun 1999: 106, fig. 12).

The suggestion of a perfumery with wedge presses is ingenious and seems likely to be correct, but it is not the only possibility. If this were a perfumery, it seems odd that no perfume vessels appeared in its active phase among the pottery published by Brun (1999: 109–15). If these were wedge presses it seems likely that they were used less (if at all) for the olive oil which served as a base for the perfume, than for processing the more precious aromatic ingredients. Clearly the facility underwent at least one change of use over the course of its short lifespan.

6.4. OLIVE CRUSHING AND OLIVE CRUSHERS IN CLASSICAL GREECE

6.4.1. The Introduction and Dissemination of Rotary Olive Crushers

The 'standard' Roman olive crusher for most modern scholars is the *trapetum*, immortalized by Cato (*RR* 20.2; Drachmann 1932) (Fig. 6.22), a rotary mill with two hemispherical millstones set in a large

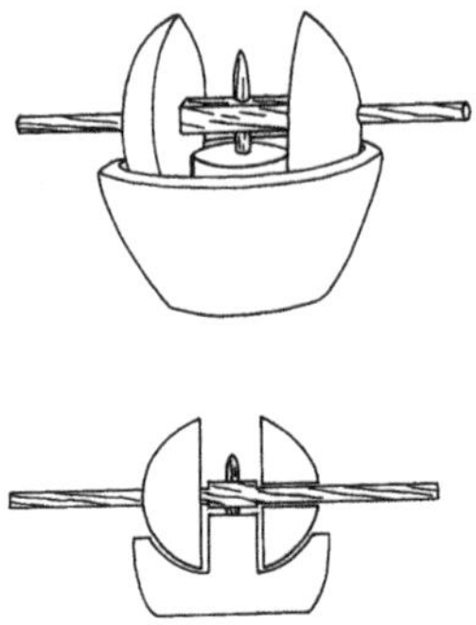

Fig. 6.22. Trapetum (H. Forbes and L. Foxhall).

stone basin (mortar). Although *trapetum* must be derived from a Greek word, *trapeo*, 'to squash up [grapes]', it is not clear when or where the machine was invented, much less when or why it came into widespread use, although it is possible that rotary mills per se are much earlier than has previously been thought (Morel 2001). The reason for taking the *trapetum* as a starting point is that, thanks to the extensive Roman evidence, one can be absolutely sure that it is genuinely an olive crusher and not primarily for some other process. However, it may not have been in common use for crushing olives in Classical and Hellenistic Greece.

The excavators of the sixth century BCE olive pressing installation at Klazomenai (see above section 6.3.2) have suggested that the site preserves the earliest example of a rotary olive crusher (Figs. 6.4; 6.7; 6.8), set in hole 1, cut into the bedrock, immediately adjacent to the wall of the structure (www.klazomenai.com/isliginikincievresi_eng.htm). At approximately 2 m in diameter, this hole is much larger than rotary crushers known from later periods. The bottom of the hole is flat, with a wide, shallow circular depression in the centre (Fig. 6.7), quite unlike the raised column to take the rotary mechanism usual on later rotary crushers. Just outside hole 1 there is a smaller hole to one side. The excavators of the site restored this as a posthole supporting a platform from which the crusher could be operated (no other postholes appear to exist). Because it is close to the wall it would have been impossible to operate it at ground level with people or animals pushing the millstone by walking around it in the usual way. The flat depression in the centre was restored as the seating for a post which served as an axle for a rotating millstone, although the central depression itself shows no wear. Hole 2 has the same kind of central depression in the bottom, but this was not restored as a crusher. The straight walls and flat floor of hole 1 are unlike virtually all later rotary crushers which are generally higher at the edges and slant downwards towards the centre (as do the curved crushing surfaces of the millstones). This design alleviates the problem of the pulp becoming clogged behind the millstone around the sides of the mortar. Although the excavators appear to have reconstructed a working crusher in hole 1, apparently using a modern millstone, it would be comforting to have more secure evidence for this reconstruction than is presently published. As there appears to be a large stone roller on the site (Fig. 6.4, sitting in hole 1), and a rock-cut crushing bed running north-south, located to the east of hole 1, it seems more likely that olives were crushed using a 'roller and bed' crusher (see below, section 6.4.2), and that hole 1 served a different function such as a settling tank.

Millstones from rotary crushers can often be roughly dated by their shape. The earliest securely dated millstones from rotary olive crushers come from Olynthos. Unfortunately, the excavators never found any complete '*trapeta*' *in situ*, but only the crushing stones, in most cases built into the walls of houses (Robinson and Graham 1938: 338). Previously I believed that the Olynthos and the Pindakas crushing stones must have come not from *trapeta* using two millstones, but from some kind of single-stone crusher (Forbes and Foxhall 1978: 42). However, one proper mortar for utilizing a pair of crushing stones does exist at Olynthos, as can be seen in Fig. 6.23 (Isager and Skydsgaard 1992: 62, pl. 3.10), showing two matched crushing stones with their mortar. Although this crusher is presently located on the site it has, of course, no excavated context, and Cahill (2001: 334, n. 49) believes that this equipment was moved onto the site from elsewhere. Though this find does show that at least some of these early rotary crushers could use two stones, it is also possible that some operated with only one millstone. The original excavators also illustrate a spouted mortar which might have been part of yet another kind of olive crusher (Robinson and Graham 1938: pl. 78, 10; cf. Frankel *et al.* 1994: 97–8).

Fig. 6.23. Trapetum and mortarium from Olynthos (Isager and Skydsgaard 1992: pl. 3.10, reproduced by permission of J.-E. Skydsgaard).

Three of the five[6] Olynthos crushers were found built into the walls of House A v 9, and a fourth was found (in the wall?) in House A v 10, next door. None of those for which dimensions are published are close enough in size to have been a pair (Table 6.1). The fact that they are built into house walls probably gives them a reasonably secure *terminus ante quem* of the mid fourth century BCE. The fifth came from the courtyard of House A 6 and is discussed in its archaeological context in section 6.3.3 above. Of all the Olynthian olive crushing stones, this is the most likely to belong to a 'working' olive press (see Fig. 6.24). It does not appear to have been part of a pair, but no mortar is present, and we cannot be absolutely certain that it was found in its primary working context.

The Olynthos crushing stones are distinct from their later Roman cousins in terms of their shape and sometimes size. The stones are flat on both surfaces, with a relatively small curved crushing surface on the perimeter edges. Compared with many of the Roman and Late Roman crushing stones found in Greece they are very large: often 0.8 m in diameter and above. As noted, they do not necessarily come

[6] The reference to a sixth crushing stone found in the court of house A1 (Robinson and Graham 1938: 208) is a phantom. A press bed was found in the court of this house (Robinson 1930: 43 and Fig. 125), which is described in the text in a rather muddled way, and by Olynthus VIII Robinson and Graham seem to have misinterpreted this description themselves.

Table 6.1. Millstones and processing installations from Olynthos (after Cahill 2001: 240).

House	Equipment
A6	Millstone from rotary olive crusher in courtyard. Twelve upper grindstones from grain mills, and an unknown number of lower stones (which do not survive). Possible press in courtyard.
A v 10	Millstone from rotary olive crusher in courtyard. (Sales inscription includes a 'pithos room').
A v 9	Three millstones from rotary olive crusher reused in foundations.
A 1	Stone press bed in courtyard.
A vi 8	Two cement platforms, room g.
A vi 10	Cement platform with gap; stone mortar in room i.
A vii 9	Cement platform in room b.
A viii 10	Cement floor, room a.
A 4	Unusual rotary millstone and installations in NW, S, SE, SW of courtyard.

in pairs like the millstones of the 'classic' Roman *trapetum*. Other crushing stones which might be of fifth–fourth century BCE date (and there are very few documented) also show these same characteristics (Forbes and Foxhall 1978: 41–2 and Fig. 4, here perhaps misidentified as being from single stone crushers). Drachmann (1932: 45–6

Fig. 6.24. Crushing stone from House A 6, Olynthos (Robinson 1930: fig. 184).

and Fig. 11) discovered one of these with no archaeological context, sitting outside the Nauplion Museum. It measured 0.605 m high, 0.13 m thick with the square hole on the curved side measuring 0.125 m per side. Drachmann recognized that the reddish grey stone of which it was made was different from the Roman crushing stones, but he could not date it and believed that it was a hemispherical stone which had been cut flat on the outer surface. Only one, from Pindakas, Chios (D. 0.82 m; Th. 0.19 m; square hole 0.12 m) (Fig. 6.25a) comes from an excavated context, dating to perhaps the mid fifth–fourth centuries BCE (Boardman 1958–9: 304), and I am inclined to believe that it belongs to the later end of this chronological range. However, the Methana survey has identified one crushing stone of this type (Fig. 6.25) from a village site which flourished in the Archaic–Classical periods, but declined during the Hellenistic

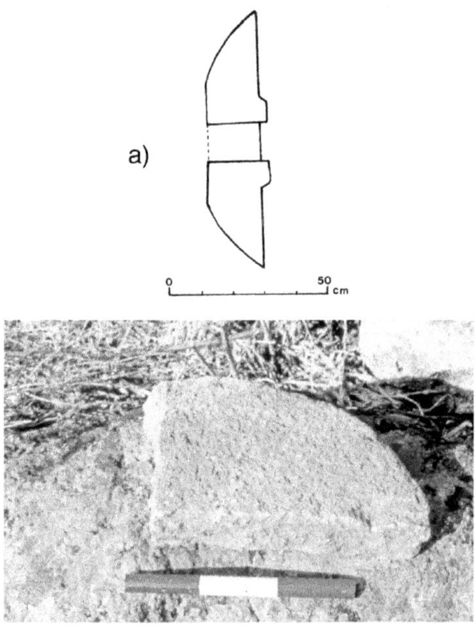

Fig. 6.25. Millstones from early rotary crushers a) Pindakas, Chios (after Boardman 1958–9: 304); b) Methana (L Foxhall; Mee and Forbes 1997: 261, fig. A1.7).

period and was not reoccupied until Late Roman times. The site history strongly supports the Classical dating of the stone (Foxhall 1997: 261, Fig. A1.7 [MS67]); the small and very rounded trapetum stones of the Late Roman period found on Methana are quite different in shape and size.

Throughout the Mediterranean, the rotary olive-crushing mill has generally been thought to be a comparative latecomer, although this view may be challenged by Morel's (2001) discovery of a possible rotary (grain) mill in a late sixth century BCE Greek context. In Palestine/Israel they do not appear before the last quarter of the fourth century BCE (Frankel *et al.* 1994: 34–5). In the Hellenistic city of Maresha (Marissa) twenty olive presses dating to the third–second centuries BCE were found. The 'lens-shaped' millstones (as the excavators describe them) of the crushers are similar in size to the fourth century BCE Greek ones (0.7–0.85 m), though they are slightly more rounded. As noted above (section 6.3.6), the closest Greek parallel may be the pair of millstones from Argilos dating to the later fourth–third century BCE. Most interesting, these crushing mills from Maresha appear to have used only a single millstone (Kloner and Sagiv 1993: 120–5; 129–30). At present it appears that the rotary olive crushers of Israel are the next earliest after those documented for fourth century Greece. The presence of early rotary crushers in both Olynthos and Argilos might suggest that the Macedonians had something to do with the invention and dissemination of this machine, and that this process was connected with the formation of the large estates of a Macedonian elite, but it would be difficult to support this hypothesis on the basis of so few early and well-dated examples. Aischines' (2.156) accusation that Demosthenes fawned upon Philip in an attempt to ransom his friends, allegedly 'captives in chains digging in Philip's vineyard', may also suggest the development of large, intensively cultivated, slave-worked farms under Macedonian control, despite the unreliability of any information from such a polemic context. The rotary crushers of Italy, Spain, and North Africa can be dated no earlier than the second century BCE, and many, of course, are later.

Although the rotary olive-crusher is recognisable in a few places over a wide area of Greece as early as the fourth century BCE, the paucity of finds suggests that is use was limited. It is possible that such early '*trapeta*' dating to the Classical period lie in wait unrecognised in survey transects or excavations. If so, I do not think there are many. For example, on the Methana survey (Foxhall 1997) there are about forty-seven press assemblages, represented by both single items and clusters of several items of pressing equipment. This includes a remarkably large number of olive crushers for a small peninsula, yet there is only one that could be securely identified as belonging to this early type of rotary crusher. Other surveys have found none at all (see below, section 6.5). Most of those which have been discovered in the course of intensive survey are of the smaller, more rounded types characteristic of the *trapetum* of the Late Roman period (see below, section 6.5). Why do press installations in general, and olive crushing equipment in particular, which can be securely dated to the Archaic, Classical and earlier Hellenistic periods, seem to be so invisible?

I think there are three main reasons for this. First, olive presses were often located in the countryside, not in towns and cities, hence they are rarely picked up by urban-based excavations. Sometimes olive and wine presses are not even located on 'farmsteads', as the number of presses found in survey transects (not on 'sites') testifies. Many of those found by the Methana survey and other surveys (Table 6.2) appear to be 'off-site' finds, though in many cases they may not be *in situ*. Occasionally on Methana, 'off-site' pressing equipment can be associated with a particular farmstead (as in the case of sites MS19 ('farmstead' site) and MS20 (small isolated processing site with the mortar of a rotary olive crusher) (Fig. 6.26). However, most of these are either Roman or later (see below), or are simply undatable, so this is clearly not the whole explanation. Next, ancient equipment for olive and wine processing, such as large, useful stone troughs and mortars, and grinding stones, are regularly removed from archaeological sites by later inhabitants. This is not surprising, especially in areas where easily workable stone, especially stone suitable for grindstones of various kinds, is in short supply. In our experience on Methana, where easily worked volcanic stone is readily available, such items as the mortars from *trapeta* are inclined to 'walk' to the nearest well or cistern, where they have often been turned into clothes-washing basins. In the Atene Survey, Lohmann (1993: 495 and pl. 69.3) found the central columns from two *mortaria* of *trapeta* on ancient sites

Table 6.2. Pressing equipment found in survey.

Methana Location	Description	Date
os–A18.2	Press weight block. Probably associated with MS1, in the same transect, or MS3.	R–LR
os–A21.2	Trapetum base. Most likely associated with MS8/9 or MS7.	R–LR
os–B23.4	Orbis of trapetum.	R–LR
os–B25.2	Press weight block.	Undatable
os–A9.1	Trapetum base.	R–LR
os–B10.2	Probable trapetum base.	R–LR?
os–E27.3	Wall with press, press bed.	Undatable
os–D28.3	2 patitiria (modern?), one with press bed perhaps older than present use.	Undatable
os–C29.2	Rough press cut in living rock (large boulder).	Probably R–LR
os–D29.2	Press weight block with localised sherd scatter.	
os–E3.2	Crude boulder-cut press bed (slides). Possibly associated with MED/T/EMOD reuse of MS211?	R–LR?
os–E8.1	Patitiri and rough press bed.	Undatable
os–C27.1	Trapetum base. Perhaps associated with MED/EMOD reuse of ancient pressing equipment on MS75 or MS113.	R–LR
MS14	Possible millstone from olive crusher.	R–LR
MS19/20	MS19, 'farmstead' with associated, but separate olive press (MS20) represented by trapetum base. Goes with press bed found in transect A5.1.	R–LR
MS22	Press bed.	R–LR
MS53	'Olive press' mentioned in site notes but no photos or further information. Holes in a wall?	Undatable
MS66	Patitiri made out of ancient bits of press, crusher.	R–LR
MS67	Section of circular millstone from C rotary olive crusher. Press weight, most likely belongs to LR 'farmstead' re-occupying earlier village site and may go with crusher found at MS66.	C, R–LR
MS69	Press weight.	R–LR
MS70/71	MS70, press bed in front of beam holes cut into large boulder.	
	Almost certainly associated with MS71 nearby.	HE
MS73	Possible press: boulder cut for beam holes.	Undatable
MS75	Roller and bed crusher; press bed; boulder weight; mortarium from trapetum.	
	Several other possible roller and bed bases.	R–LR, reused MED
MS100	Fragmentary trapetum orbis.	R–LR
MS101	Two trapetum bases; two small fragments grey lava trapeta; rough bowl, base of olive crusher, cut from boulder; fragmentary bowl, base of olive crusher.	
	Trapetum base, transect C20.2, probably from this site, present location is probably tertiary use.	R–LR, reused MED

(*Continued*)

Table 6.2. (*Continued*)

Methana Location	Description	Date
MS106	Large press bed; block with hole for press beam; fragment of mortarium-type base from olive crusher; fragmentary stone trough; press weight block.	HE
MS109	Trapetum base and two orbes; press bed; block with round hole, possibly for press beam; rough stone mortar.	R–LR
MS113	Rough press bed cut from a boulder.	Undatable, probably EMOD
MS114	Press weight block; trapetum base; frag, trapetum base; stone roller, probably from a crusher.	R–LR reused and supplemented MED/EMOD
MS115	Part of press/recepticle; probable trapetum orbis.	R–LR
MS116	Block with 2 holes to take press beam; probable fragment of trapetum orbis.	
	Probably associated with beautiful press nearby, transect C29.1.	R–LR
MS120	Block cut with holes to take beam; roller.	Undatable, C–LR
MS121	Large rectangular block with two holes for press beam.	C–HE?
MS122	Press bed, press weight, block with hole for beam. May go with patitiri in transect C1.1.	R–LR
MS 123	8 patitiria, 1 or more nicely cut from natural boulder, including 6 press beds.	Interpreted as C–HE, LR and EMOD agricultural site.
MS123B	2 circular cuttings in boulder to take press beam.	Associated with MS123.
MS209	Boulder cut for press beam, base of olive crusher, 2 fragments of press bed.	R–LR
MS210	Press bed.	R–LR
MS211	Large cut block with beam holes; cut block with beam hole, possibly reused as press bed; good cut blocks, some reused as press weights; fragment of trapetum base. Probably associated with probable patitiri, transect E3.1.	R–LR
MS216	Trapetum base; stone with round hole to take press beam.	
	Press bed found in deserted EMOD village of Panayitsa may belong with this site.	R–LR
MS218	Trapetum base; press bed.	R–LR, with MED/MOD reuse.

Southern Argolid Location	Description	Date
A8	Press bed; possible roller.	LR
A60	Press bed.	C–HE

Table 6.2. (*Continued*)

Southern Argolid Location	Description	Date
A61	Ceramic basin sunk in cement floor; press bed; possible rollers.	C–HE
B5	Trapetum orbis.	R–LR, found in church.
B6	Press bed.	undatable, built into modern terrace wall.
B78	Press weight block; press bed; 2 fragments of trapetum base.	HE–LR
B91	Trapetum base; possible trapetum fragment.	R–LR
B103	Press weight block.	Undatable, C–LR
C11	Press bed; trapetum fragment.	Probably both R–LR
D8	Press bed.	Probably ancient but undatable
E7	Press weight block; trapetum base.	R–LR Dug up together during house building.
E12	Trapetum base.	LR
E26	Press bed; press weight block; trapetum orbis.	LR
E30	Press weight block.	C–HE
E38	2 press beds; press weight block.	C–HE
E45	Trapetum base.	R–LR
E50	Press weight block.	Undatable, found in church.
E52	Press weight block, possible roller, possible press weight block.	C–HE
E54	Press bed; possible press weight block.	C–HE
E70	Two press weight blocks.	Undatable, C–LR or later.
E81	Press weight block.	Undatable, reused in MED building.
os–F2	LR 'farmstead', trapetum base found reused as watering basin at spring nearby.	R–LR
G1	Boulder weight.	Undatable, LG–MED
G12	Trapetum base.	R–LR
G14	Press bed; press weight block.	Undatable

Kea Location	Description	Date
Site 9	Press bed.	C–HE
Site 15	Two press beds.	Probably HE–LR
Site 64	Millstone from rotary crusher.	Probably HE–R

Table 6.20. (*Continued*)

Lakonia Location	Description	Date
os – J219	Press bed(?), about 40 m from 'farmstead' site.	C
Q359	Possible orbis from trapetum.	R–LR

Atene Location	Description	Date
CH31	Press bed.	Perhaps C–HE
LE16–LE17	LE16: trapetum base fragment; LE17: press bed spout, press weight block.	R–LR
AN3	Trapetum base fragment.	R–LR

Aetolia Location	Description	Date
Tolofon (E)	666–237 'large fragment of ancient oil press'.	Undatable.

Melos
None recorded

Minnesota Messenia Expedition
None recorded
os = off site

(AN3, LE16), but the *mortarium* bowls themselves had disappeared, suggesting that they suffered a similar fate to many of the Methana *mortaria.* Similarly, the one millstone from a rotary crusher found at Delos (Brun and Brunet 1997: 597, Fig. 18; Brun 1999: 149) was found as part of the mole in the harbour and has no primary archaeological context. Its present condition is very worn, but it appears similar to the Late Roman types. This phenomenon (rather than an absence of olive trees) may well at least partially explain the paucity of olive crushing equipment dating to any period found on the Boeotia Survey (J. Bintliff and A. Snodgrass, pers. comm.). Last, probably the most significant explanatory factor is that even in the fourth century BCE, and certainly earlier, people crushed olives by other means than using rotary crushers. It is clear from the Roman agronomic writers that centuries after the invention of the *trapetum* other devices and methods were also used to crush olives. The same phenomenon is well-documented in Palestine/Israel: older techniques survive even when 'improved' tech-

a)

b)

Fig. 6.26. Methana: a) MS19, 'farmstead' site and b) MS20, isolated agricultural processing site with olive crusher (L. Foxhall).

nologies have been introduced (Frankel *et al.* 1994: 28, 31). Grain mills provide a useful analogy here. At Olynthos (Table 6.1) and other sites both hopper-rubber mills and saddle querns were used concurrently, because the more elaborate and expensive mills are either too expensive, or not suitable for other reasons in particular social, economic, or technical contexts. The same must have been the case with olive crushing. Simpler and archaeologically less-identifiable or visible equipment must often have been used.

If this last factor is indeed the most important, then the implications are significant indeed. In particular, it suggests that, broadly speaking, the scale of olive processing was relatively small and the degree of specialization of processing installations considerably limited before the later Hellenistic and Roman periods in the Greek world.

6.4.2. Other Ways of Crushing Olives

Many types of olive crushers are documented in the ethnographic and archaeological records from all over the Mediterranean (Champs-Fabrer 1953; Amouretti *et al.* 1984; Amouretti 1986; Sordinas 1971; Forbes and Foxhall 1978; Benaki Museum 1978; Frankel *et al.* 1994; Amouretti and Brun 1993; Galili *et al.* 1997).

Some of the olive crushers documented in more recent ethnographic and historical settings are based on the same principle as the *trapetum* or *trapetum*-like rotary olive crusher. Many look very like English apple crushers for cider-making, with one or more millstones, most often cylindrical in shape, rolling around on a concave stone surface or shallow stone bowl (Sordinas 1971: pl. 2; Benaki Museum 1978: pl. 75). For almost all of these, including the *trapetum*, the crushing stones would have to be made by specialist stonemasons. To judge from the volcanic stone often (though not always) used in their manufacture, the stone itself had to be imported in many areas. Where stone had to be imported, it may sometimes have been in the form of pre-cut, roughed-out millstones.

The other crucial factor in terms of expense is the number of metal fittings which had to be provided and/or installed by specialist craftsmen. This might be enough to put some crushers (including *trapeta* and other rotary crushers) out of the reach of even many better off farmers, or at least make them uneconomical (Sordinas 1971: 11). For example, Sordinas (1971: 8–11, pl. 2) documents the use of a cumbersome, comparatively inefficient, single-stone crusher for at least a century after the introduction of an improved crusher utilizing several smaller stones in eighteenth-nineteenth century Kerkyra. This is because the former could be made by village craftsmen out of local wood and stone, with the only metal parts being one pin and a few large nails, while the latter could only be produced and assembled by specialist iron workers and stonemasons who were urban-based (incidentally also increasing transport cost, as well as the costs of labour and materials). In antiquity, the substantial cost of a specialist, highly-crafted olive crusher might well be thought an unnecessary expense, particularly in mixed farming regimes, where few farmers had very large numbers of olives. The reason the improved crusher caught on in nineteenth-century Kerkyra at all was that there was a considerable amount of relatively specialized olive growing at that time on the island (Sordinas 1971: 1–2). The fact that in most areas an olive crusher would only be used every other year, at most, could make such equipment even more uneconomical. It also explains why, in contrast to professional grain milling establishments,[7] large-scale olive pressing establishments rarely seem to have been worthwhile as a business on their own. Grain was ground throughout the year, and there was always a steady, if not very large, demand for the service. Olives, however, are only processed at the time of the harvest, once every two years. It is perhaps not coincidental that Theophrastos (*CP* 1.20.4) notes that in the vicinity of Olynthos the olive is less inclined to biennial fruiting, and that this is precisely the area in which there is the best evidence for the early use of the rotary crusher. Large, expensive, specialized crushers are only good value if there is specialist olive growing and oil processing on a sizeable scale as well. As has already been argued, this was not the case for most of Classical Greece.

In Classical Greece, even on the holdings of wealthy land owners, simpler and cheaper devices must regularly have been used for olive crushing. The simplest of all was the mortar and pestle. However, crushing olives with even a large stone mortar would have been slow and arduous, consuming considerable amounts of human labour. Within a pressing installation a storage container for batches of crushed pulp would have been essential to accumulate enough to fill the press.

Another method, still in use in many parts of Greece in the recent past, is the 'roller and bed' type crusher. Here the operator rolls a cylindrical, column-drum-shaped stone roller repeatedly over the olives spread out over a stone base (Forbes and Foxhall 1978: 39–41; Sordinas 1971: 8; Frankel *et al.* 1994: 30, Fig. 19 and 31, 98, Fig. 95 and 99). The olive crusher of this type, illustrated in Fig. 6.27, is from a Medieval settlement on Methana. When we discovered it, our archaeological guard (also a practising farmer) immediately recognized what it was, and is shown here giving a demonstration of how it was used 'in the old days'. As with a mortar and pestle, such crushers are adequate as long as large amounts of olives do not need

[7] Even they are less common in classical Greece than in the Roman world, since most grain grinding seems to have been done within the household.

Fig. 6.27. 'Roller-and-bed' type olive crusher from Mediaeval settlement site, Methana (L. Foxhall).

to be processed for pressing quickly. As long as sufficient labour was available, it was still probably cheaper to operate several crushers of this sort than to invest in a rotary crusher. Little more was needed than a redundant column drum (possibly also in use as a roller for packing down mud roofs or threshing floors) and a large, flat stone. And, at the end of the season, the apparatus, such as it was, could be dismantled.

A serious misunderstanding about the technology of olive crushing which is periodically resuscitated (Amouretti 1986: 162 and Frankel 1994: 31, 78; White 1967: 227; Brun 2000: 283; 2004: 8) is the idea that olives were trodden, like grapes, before being pressed. Although this might be theoretically possible, it would be immensely time-consuming and extraordinarily ineffective, since it would do little more than bruise the olives. The only process for which this might be potentially useful is for making particular types of table olives, particularly from hard green olives, where the unripe fruits are sometimes cracked or bruised before pickling in brine (nowadays this is done by squeezing the whole fruits gently and briefly in a press). For the ancient Greek world, there is no evidence that such a method was ever used. The misunderstanding seems to have arisen from White's (1967: 227) interpretation of Columella's (*RR* 12.52.6–7) olive crusher, the *canalis et solea*, literally the 'channel and shoe'. White took this quite literally and imagined that it consisted of a set of special shoes worn for treading olives. He supported this argument by reference to a Roman relief in the Palazzo Rondanini (Fig. 6.28) which he thought showed a cupid treading olives. In fact, the Rondanini relief actually shows olives in a heap in a special

Fig. 6.28. The Rondanini relief (Roman) (Drachmann 1932: 144, fig. 10).

area waiting to be crushed, since there is a perfectly good rotary olive crusher shown in operation on right of this relief. The cupid in question is standing *behind* the heap of olives. The term 'shoe' in the *canalis et solea* was probably not literally a shoe, but more likely refers to the crushing stone either of a single-stone rotary crusher or the roller of a 'roller-and-bed' type crusher described above.

Amouretti (1986: 162) has tried to support the notion of 'olive treading' with reference to the lexicographers of late antiquity. The word *kroupezai* means 'a device for crushing olives', but it also is used to refer to a type of high wooden shoes said to be typically worn by Boiotians. The lexicographers (Hesychius, Photius, s.v. *kroupezai*) seem to have got these two distinct meanings conflated. Pollux (7.87) mentions them as shoes, but significantly also mentions the word again (10), this time in a singular, diminutive form (*kropezion*) in a section of the *Onomastikon* which deals largely with furnishings and equipment.[8] This latter use sounds rather odd for shoes (only one?), but makes reasonable sense for a small olive crusher or crushing stone. I am inclined to think that the word primarily meant 'olive crushing stone', whether a roller or rotary millstone, and was then applied to heavy, cloggy shoes, perhaps as a joke.[9] *Peze* means 'foot' but in a very broad range of meanings (including 'bottom, base'), and the verb *krouo* means 'to smash, pound, beat'. The term could easily be understood as a 'roller-and-bed' type olive crusher.

[8] This section of Pollux incidentally includes many references to equipment in the Attic Stelae, see Pritchett 1953: 209, and the appendix by Pippin, 318–28.

[9] Analogous to the old English slang terms 'beetle-crushers' or 'winkle-pickers' for 1960s styles of shoes? Many of the references to *kroupezai* are in fragments of comedy (Pritchett 1953: 209). What better way to poke fun at rustic Boiotians than to have them wearing 'olive-crushers' on their feet?

6.5. PRESSING EQUIPMENT IN THE COUNTRYSIDE: FINDS FROM ARCHAEOLOGICAL SURVEY

6.5.1. General Issues

Over the past thirty years, intensive archaeological field survey has become increasingly important as a technique for studying agriculture and land use in Greece and other regions of the Mediterranean over the long-term. Significant amounts of pressing equipment have been systematically recorded by a number of survey projects (Table 6.2), and it has now become feasible to use this data comparatively to explore the changing patterns of exploitation of the olive and the vine across a wide span of ancient Greek time and space. As will become evident in the following analysis, survey data has its limitations. Dating can rarely be as precise as excavated contexts would allow. There are also considerable discrepancies in levels of observation and recording between surveys: the most recent are generally the most detailed, as the methodology of intensive survey has become more refined with the collective experiences of surveyors. Nonetheless, this material is tremendously informative about the scale and organization of wine and oil production on a regional level. Moreover, it allows us to see the production systems of Classical times in a fully comparative temporal perspective. In this section it will be necessary to expand the chronological focus to explore the later Hellenistic and Roman pressing equipment in and beyond these Greek regional contexts as well.

Several of the published Greek surveys have recorded little or no pressing equipment (Table 6.2). Notably, these tend to be the earlier ones, e.g. Melos, in which none is mentioned at all (Renfrew and Wagstaff 1982), or the less intensive ones, e.g. Aetolia, which notes only one: 'a large fragment of an ancient oil press' on site Tolofon E (666–237), dated to the Classical–Hellenistic period on the basis of the pottery scatter examined at the site (Bommeljé *et al.* 1987: 110). It is difficult to be sure whether the lack of pressing equipment is because it was not there or because it was not observed or recorded. This second example highlights a related problem: when it comes to pressing equipment, many surveyors did not know what they were seeing and hence failed to describe it clearly enough to be of use. It is impossible to determine from the publication whether the Tolofon example is a press bed, a weight block, part of a rotary crusher, or something else altogether. Despite these potential problems of data collection, it is clear that in many parts of Greece very little pressing equipment of any period is discovered on intensive survey; for example Lohmann's (1993: 374, 513, sites LE16–LE17, CH31, AN3, AN23, and possibly PH2) survey of Atene (SW Attica) found only seven items on six sites. In contrast, on the Methana (Mee and Forbes 1996) and Southern Argolid (Jameson *et al.* 1994) surveys, considerably more pressing equipment was evident than elsewhere, and these cases will be discussed in detail below.

Both van Andel and Runnels (1987, in the context of the Southern Argolid survey) and Lohmann (1993, for Southwest Attica) have argued that olive cultivation on a substantial scale was important in the regional economies of their respective regions. Runnels (Runnels and van Andel 1987; van Andel and Runnels 1987; cf. Acheson 1997) has argued that in periods of high population and political stability in the Southern Argolid, the soils and climate of the area dictated that agricultural intensification generally took the form of increased investment in olive trees with the aim of producing olive oil for export from the region as an income generating commodity. In the Late Classical–Early Hellenistic period (ca. 350–250 BCE), he argues, this process is evident in the proliferation of small, scattered farmsteads worked by relatively well-off farmers with an 'orientation to olive culture' (Jameson *et al.* 1994: 384–91) Lohmann (1993), studying a particularly dry and rugged part of southwestern Attica has taken a similarly line of argument. His hypothesis that the small, scattered farmsteads of the fifth and fourth centuries BCE were primarily geared to grain and olive production is largely based on the finds of undatable terrace systems in their vicinity, which, he asserts, were used in classical antiquity for growing olives, since allegedly nothing else would have grown there (Lohmann 1993: 196–219). I do not believe that either of these models holds up under close scrutiny.

When it comes to the analysis of pressing equipment found in the course of intensive survey, three major problems arise 1) determining its functions, 2) dating it, and 3) determining its life span.

First, virtually all pressing equipment found by survey archaeologists is identified with the processing of olives.[10] I have argued above that in Classical Greece presses were regularly multi-purpose installations which could be used for the manufacture of wine as well as oil, and for other industrial processes as well. The existence of a press on a site is no guarantee that olives were processed there, or that the press was used exclusively for making oil. Indeed, much pressing equipment has been found 'off-site' in the course of survey (Table 6.2), exacerbating the problems of dating.

Next, without a stratified context, most pressing equipment is very difficult to date, and much remains undatable. Often, surveyors are too eager to date pressing equipment to their 'preferred' period when there is little archaeological justification for so doing. For installations on a surveyed site which was occupied only for a short time or during limited periods, sometimes the balance of the evidence favours one period over another, though a definitive date may still be elusive. However, with press beds, troughs and basins, rock-cut receptacles below press beds, and holes cut to take press beams, 'stylistic' traits may indicate a 'regional' style persisting for a long period of time which therefore has little chronological significance for the dating of any particular artefact.[11] However, simultaneously and confusingly, different 'styles' and types of pressing equipment may be in use in the same area concurrently, depending on the social, economic, and technological context of the particular pressing establishment.[12]

[10] This is the case even with very careful surveyors: e.g. on Kea the press beds at site 15 are described as: 'rock cut olive presses' (Cherry *et al.* 1991: 296, fig. 13.9 caption); 'two spouted presses for olives' (Cherry *et al* 1991: 83); the press bed on site 9 is called 'a rectangular stone press bed for olives' (Cherry *et al.* 1991: 78).

[11] On the range of types in N. Africa, see, Mattingly and Hitchner 1993: 453–4. Cf. Mattingly 1996a, on the regional differences in techniques for attaching the fixed end of the press beam in lever and weight presses; Frankel in Frankel *et al.* 1994: 40–5, on the differences in lever and weights presses between northern and southern Israel.

[12] Frankel and Avitsur in Frankel *et al.* 1994 have amply documented the simultaneous existence of many different press types in both ancient and modern times in Palestine. For ethnohistorical documentation of the same phenomenon see Sordinas 1971 (Corfu) and Cassanova 1993 (Corsica). The larger economic implications of alternative regional 'preferences' in pressing technologies are taken up by Mattingly (1996a).

Despite the health warnings, there are a few chronological footholds that seem to be reasonably reliable. In Greece rotary crushers are usually roughly datable. As argued above, proper *trapeta* are virtually always Roman or Late Roman, while rotary crushers with distinctive flattened stones are fourth century BCE or a bit later. Other possible chronological indicators are present in some counterweight stones, which Brun (2004a:11) has carefully typologized. Round counterweights for screw presses do not appear before the Roman period, and even then they are rare in Greece. Brun's claim that his 'type 41' counterweight block, with two rectangular holes in the upper surface, is characteristic of Classical and Hellenistic installations seems to be generally true, though there may be some exceptions (see below).

Last, closely related to the second point is that sturdy, stone pressing equipment may have a very long functional lifespan. This need not necessarily be continuous: there are now well-documented examples of pressing equipment of one period being later reused, sometimes for pressing, but sometimes for other functions (Methana: Foxhall 1996, appendix 1 in Forbes and Mee 1996; Maresha: Kloner and Sagiv 1993: 120; Southern Argolid: Jameson *et al.* 1994: sites B91, E81). As just noted, longevity is a particular problem in the case of press weight blocks. Later inhabitants may be keener to walk off with a press bed for a DIY wine treading floor, or remove the central column from a *trapetum* bowl to make a handy watering trough, than to adapt a press weight as anything other than what it was intended for, with the exception of reuse as building material. Such blocks can often be reused for later presses with little or no modification. However, it is clear from the many which are found modified by the addition of extra mortise holes (often at the ends) that these blocks were regularly adapted for different kinds of press mechanisms in later periods. Also, while press beds may be rendered dysfunctional by chipping or cracking, and trapetum bowls may be conveniently adapted as containers or troughs, press weight blocks are durable and long lasting in comparison, and much more difficult to move. The other item less prone to later reuse (other than as convenient building stone) is the hemispherical crushing stone of the *trapetum*, the *orbis*. Once the *trapetum* was no longer a standard type of olive crusher and there was no longer the skill available locally to maintain the intricate metal and wooden mechanism, the millstones became virtually useless and were not easily adapted to other kinds of grinding or crushing machinery, thus providing good evidence for archaeologists.[13]

On Methana and in the Southern Argolid particularly, large amounts of pressing equipment were found in comparison with most surveys (Table 6.2). On the Atene survey less pressing equipment was found, but was important to the Lohmann's (1993) interpretation of the local agricultural landscape. Detailed analysis of the finds from these three surveys reveals a number of interesting trends over the Classical through Late Roman periods. I have therefore used these as case studies to investigate changes in how olive and vine cultivation fitted into regional agricultural systems over the long-term. In chapter 1.4 I explored the way in which the olive has regularly been represented by scholars as a 'transformational' crop; that is, either the cause or the result of profound social, political, and economic change. The data presented here underpins the views expressed earlier: that technological change is more likely to be a symptom than a cause of social and political change.

6.5.2. The Methana Survey

For such a small survey area, the amount of pressing equipment which has survived on Methana must be almost unparalleled in Greece: about forty seven press assemblages were found. Very little can be assigned with certainty to the Classical and Hellenistic periods: only five sites with occupation from these periods produced any pressing equipment at all, and, of these, three are certain and two are more doubtful (Foxhall in Mee and Forbes 1997: app. 1). The first of the securely identified ones is from the Early Iron Age–Early Hellenistic village site at Ogha (MS67). This is the millstone from an olive crusher which parallels the rotary crushers using millstones with two flat faces from Olynthos and Pindakas (Fig. 6.25). It is noteworthy that it was

[13] The instance in Kea of an *orbis* in use at present as a cistern cover (Cherry *et al.* 1991: no. 64.8, 156, Fig. 5.34), or the Delos *orbis* built into the mole of the harbour (Brun and Brunet 1997: 597, Fig. 18; Brun 1999: 149) are cases in point.

not found between sites or on a 'farmstead' site but as part of a 'village' assemblage, perhaps associated with a house with a tower (cf. S. Morris and Papadopoulos 2005: 169 and n.60).

The second secure example is site MS70. There is no sherd material associated with the olive press at MS70, and only a few cut blocks and some tile. The press site is almost certainly associated with MS71 nearby, which is clearly Hellenistic in date and produced no Classical, Roman or Late Roman pottery. These two sites offer a good example of a press geographically separated from the 'farmstead' site, like MS19 and MS20 from the Roman–Late Roman period. The press itself (MS70) consists of a boulder cut to take a press beam with a rock-cut press bed. The location and terrain of this 'farmstead' and its press might suggest they primarily processed olives. But there is no associated olive crushing equipment, nor is there any evidence for the separating facilities needed for making oil. It is thus impossible to determine with certainty the functions of this particular press.

The third example, MS106, is similar to MS 70/71 in scale and setting. The site is located on a high ridge above Vromolimni. There are numerous cut blocks, which seem to be mainly olive pressing equipment. There is no certain Classical pottery, but there is a reasonable amount of Hellenistic material. Helpfully, there is no Roman or Late Roman pottery, hence the interpretation as a Hellenistic 'farmstead'. The press equipment includes: a large press bed in grey volcanic lava, with a 'rusticated' chiselled interior; a stone block with a hole for a press beam; most importantly, a rim fragment of a bowl from an olive crusher; a fragmentary stone trough; and various pieces of worked volcanic stone whose function is not now identifiable. The bowl must almost certainly be a *mortarium*-type bowl, from a *trapetum* or *trapetum*-like olive crusher, though given its fragmentary state it is impossible to determine whether it has a central column and what kind and how many or what type of millstone(s) it took. This and the crusher from Ogha (MS67) are the only Classical or Hellenistic olive crushers we discovered on Methana. The one on MS106 is the earliest to be found on a 'farmstead' site.

The fourth and fifth, both less secure, examples are presses which appear on MS120 and MS121. Both are small sites located up in the mountains at Makrongu. MS120 is one of only two 'farmstead' sites which appears to be continuously in use over a long period, Classical-Late Roman, and is the site of a church in the Medieval period. MS121 is largely Classical-Hellenistic, but there is a little bit of Medieval and Turkish period material as well. The pressing equipment at each site consists of a block with holes cut to take a press beam. There are also two possible rollers on MS120 which might have been used for olive crushing, but might have had other purposes (such as compacting threshing floors or mud roofs) as well. Both sites are located very close to areas which are especially suitable for vine cultivation and are above the modern altitude limit of olive cultivation. If these presses were used for olives, the fruit must have been brought uphill to it—a situation not unparalleled on Methana—but it is not clear whether this was the case here or not. Given the unusually long period of occupation of MS120, it is impossible to pinpoint the periods during which this press was in use, nor is it possible to determine with certainty whether it was used for olives or grapes or both. Similarly, although the press on MS121 may be Classical, typologically it is undatable. It might just as easily date to some later historical period when the site could have been seasonally used at vintage time.

Almost all of the Roman–Late Roman and Late Roman 'farmstead' sites are associated with pressing equipment for both olive and grape pressing: twenty five examples are reasonably securely datable. The density of pressing equipment is of an order of magnitude similar to the very high densities of presses found in North Africa and Spain, well known for oil production for export in Roman times. On Methana the density of Roman–Late Roman pressing installations was about 1 per 3.2 km^2. In comparison, the Kasserine survey revealed a density of 1 press per 2.75 km^2; while for the whole Sbeitla–Kasserine–Thelepte region of central Tunisia the density was 1 press per 4.3 km (Hitchner and Mattingly 1993: 441; Hitchner 1993: 502). In the Djebel, west of Lepcis Magna, the density was 1 press per 2 km (Mattingly 1988a; Hitchner 1993: 502). On Methana, 89% of Roman rural 'farmstead' sites have associated pressing equipment (this does not include presses located on major settlements, e.g. MS10, the ancient polis of Methana). In contrast, in the Guadalquivir valley in Spain, 161 out of 1500 Roman rural sites (11 per cent) have pressing equipment (Mattingly 1988a; Hitchner 1993: 502).

The parts of these presses which we have normally found are: 1) stone press beds; 2) cut blocks or natural boulders with two or more circular holes to take a wooden press beam; 3) stone boulder weights with a suspension hole; and 4) stone press-weight blocks cut to take a capstan/windlass. The consistency of the diameters of holes to take press beams is significant. They range in size from 15 to 27 cm (all but 2 are in the 15–18 cm range), and can hardly have been massive beams: the small diameter suggests they were also fairly short, like the beam of the Halieis press in the Industrial Terrace (see above, section 6.3.3). The simplest explanation is that large heavy timber was an expensive commodity in short supply, so that builders of presses made do with the minimum serviceable for the job. Nor are the holes very deep: generally they range from 9–19 cm in depth. Even assuming that beams were firmly wedged in place when the press was in operation, if too much pressure had been exerted the beam would have popped out of these shallow holes. This adds to the impression that these were quite small presses which did not exert a great deal of pressure compared with the massive presses of Roman North Africa or the 'traditional' presses of nineteenth–early twentieth century Greece or Palestine (see below, section 6.8.2).

On Methana, eight examples of weight stones associated with ancient pressing equipment were found. Although there are several different types, in all cases where they can be reasonably linked to a site and/or pressing assemblage, the contexts appear to be Roman–Late Roman. Rectangular weight stones with two rectangular cuttings on the top face of the type most regularly found in the Southern Argolid (Brun's 'type 41', see below, section 6.6.3) appear, for example on MS211 (Fig. 6.29) where it is associated with a fragment of a *trapetum* bowl. Though this site seems to have had both a Classical and a Roman–Late Roman phase, the latter appears to be much more extensive and substantial and it is likely that this weight block was in use in the later period, even though it might have originated in the earlier period. Some are nearly square, with a hole going straight through, as on D29.2, associated with MS117 (Fig. 6.30), a Late Roman site with no Classical material; or the weight block on MS122, which is predominantly Roman–Late Roman. Others are T-shaped, as on MS114, a Roman–Late Roman site with no Classical material (Fig. 6.31), or MS67 which has both Classical and

a)

b)

c)

Fig. 6.29. Methana, MS211: a) site view; b) rectangular counterweight block with rectangular cuttings; c) fragment of *trapetum* bowl (L. Foxhall).

Fig. 6.30. Methana, Transect D29.2, associated with MS117, squarish press weight block with through hole (L. Foxhall).

Fig. 6.31. Methana, MS114, T-shaped press weight block (L. Foxhall).

Fig. 6.32. Methana, MS67, T-shaped press weight block showing rectangular holes straight through (L. Foxhall).

Roman–Late Roman olive crushers (Fig. 6.32); the weight could have been in use in either or both periods, though its present shape suggests it belongs to the later phase of use (Brun 2004a: 17; 1986). Whatever their date of manufacture it is fairly certain that all of these were *in use* in the Roman–Late Roman period, and the balance of the evidence suggests that most were probably made in this period.

Not all press assemblages seem to have weight stones associated with them. The clearest example is MS109 (Fig. 6.33) where all the stone parts of an olive and wine press were found except for a weight stone. If there had been a weight stone it is surprising that we did not find it. This might indicate that a lever and weights press operated by a windlass was not used here. The alternative possibilities include a lever and weights press using sacks full of stones, as is perhaps represented on the much earlier, archaic-period Boston skyphos, or that the press was operated by a screw mechanism made entirely of wood. The former explanation is perhaps more likely, since the cylindrical weight stones with central cylindrical holes characteristic of the screw press typically found in Palestine/Israel do occasionally appear in Greece in late antiquity (Forbes and Foxhall 1978: 43, Figs. 10, 11), and no such stone was found here.

It is striking that in three cases (MS209, MS211; MS109) *trapeta*, some of the equipment most clearly intended for olive processing, are located above the modern line of olive cultivation. In the absence of any evidence for climatic change, the simplest explanation is that these farms controlled land at lower altitudes where olives were grown, though they were processed back at the main farmstead. The case of MS109 is particularly clear: the site includes a structure and very high quality pressing equipment located next to a small volcanic basin suitable for vines, in which they were growing in the 1980s (Fig. 6.33). The choice of habitation next to the vines rather near the olives may have been motivated by considerations of both workload and security: vines are more labour intensive, but they are also a high-value crop and much more easily damaged by clumsy humans or careless sheep than olive trees. A watchful eye nearby might have been felt to be essential. Sometimes, however, as in the case of MS19 and MS20, the olive press is separated from the main 'farmstead' site. These examples suggest that the landholdings which belonged to a particular farm were fragmented, scattered holdings,

a)

b)

Fig. 6.33. Methana, MS109, a) site; b) assemblage of pressing equipment (L. Foxhall).

a phenomenon which has been well-documented for both the Classical period in Greece (Jameson 1977: 130–1; Osborne 1987: 37–40; 1985: 60–3), and the modern period on Methana (Forbes 1982), and elsewhere in Greece.

Many of the sites with presses and wine treading floors are located in areas which are even now considered especially good for vine cultivation. This is especially true of the sites near small volcanic basins in the mountains, such as those near Makrongu (MS120, 121, 122, 123) and Stravolongos (MS116, TC29.1). MS123 is a particularly interesting example. Here there are eight wine treading floors, several cut from natural boulders, and press beds. They almost certainly date from a number of different periods, though perhaps the majority, especially those with the best workmanship, are likely to be Roman–Late Roman, since they are morphologically close to many of the *trapeta* for Roman–Late Roman sites. The pottery on MS123 ranges in date from one Archaic sherd, through Classical, Hellenistic, Late Roman and Medieval. It is not far from the Classical–Late Roman 'farmstead' site of MS120. However, there is no indication that any of these installations were utilised for processing olives. More likely, they were designed for wine making.

It is difficult to be certain how many of the pressing installations we discovered on Methana were used for processing both olives and grapes. Nonetheless, it is likely that many if not most produced both wine and oil. The only really certain evidence that a pressing installation was used for olives is the presence of crushing equipment, such as *trapeta* (Forbes and Foxhall 1978). Given the number of *trapeta* found on Methana in conjunction with other pressing equipment, a considerable number of the twenty five Roman–Late Roman pressing installations were certainly used for olive pressing, and quite likely wine pressing as well in many cases (as noted above, many were located near vine growing areas).

It is difficult to estimate the output of oil or wine production for Roman–Late Roman Methana. Using a 'rule of thumb' ratio of ca. 500 trees per press,[14] this suggests that there was a minimum of 10,000

[14] This is the Venetian period ratio for the Southern Argolid, Forbes 1993: 217. It may represent a system of agricultural exploitation which was more 'capital intensive' than that of the Roman–Late Roman period on Methana, where it is likely that there were more trees per press.

olive trees in cultivation on Methana in the Late Roman period, and most likely considerably more. Even if each tree produced only 2 kg of oil per olive harvest (every two years),[15] this would still amount to 20,000 kg (20 mt) oil per olive harvest. Obviously these figures are very rough, and can be taken as little more than a ballpark estimate representing a bare minimum: the amount is likely to have been considerably higher. Nonetheless, they clearly highlight the special nature of the exploitation of the Methana landscape in the Roman and Late Roman period. Oil production at this scale can hardly have been solely for domestic consumption, especially given the low population inhabiting the isolated rural sites, and must have been aimed at a wider market (cf. Hitchner 1993; Mattingly 1993; 1988). The most obvious pathway for its disposal in this period must be along the well-established trade routes between the Saronic Gulf area and the Hellespont, toward Constantinople, the political and economic centre of a newly reconstituted Roman Empire. It is possible that the export of oil (and wine) from Methana was linked into a larger system of exports out of the Southern Argolid (van Andel and Runnels 1987: 113). Certainly it is a problem to determine what were the containers in which Methana oil and wine travelled in this period, though kiln sites producing Late Roman transport amphorae, among other wares, are documented by the S Argolid survey (Jameson *et al.* 1994: 402).

6.5.3. The Southern Argolid Survey

After Methana, the Southern Argolid has the highest density of presses of any survey area in Greece: twenty overall, not counting those from the excavated urban site of Halieis (discussed above, section 6.3.2). It is probable the countryside close to Halieis was exploited directly by inhabitants of the city (Acheson 1997: 172–3). For the Roman period

[15] Again, this is a minimum estimate, based on nineteenth and twentieth century oil production figures, see below. Olive trees on Roman–Late Roman Methana may have been more intensively cultivated than those in nineteenth and twentieth century Greece. Therefore, despite the technological constraints of the equipment used, it is possible that oil:tree ratios were higher in the Roman–Late Roman period, perhaps as much as double that used here, cf. Mattingly 1988a.

in the Southern Argolid territory, this gives a density of 1 press per 2.56 km with an overall density of 1:2.56 (Fig. 6.33, cf. Methana overall levels of 1:67). However, the presentation and dating of this material is problematic. Instead of grouping all the oil/wine processing equipment found on any site together, the press beds and press weights are presented with the discussion of Classical material, while the crushers are presented with the Roman and Late Roman material (Runnels in Jameson *et al.* 1994: 384–93, especially table 6.6, 398–400, 402 especially table 6.9). This disguises the fact that a number of the weight blocks and presses dubbed 'Classical' were in fact found with Roman or Late Roman crushing equipment, as part of the same assemblage, and are therefore probably not Classical in date.

In the Southern Argolid survey, press weight blocks normally seem to have been dated on stylistic grounds: the rectangular blocks with two holes on one face (Brun's 'type 41') are regarded as Classical no matter what the context, though no argument supporting this stylistic dating is presented in the publication, and no account is taken of the use/life of the equipment. As explained above (section 6.5.1), the relationship of 'style' to chronology is complex. In fact, of the two published weight blocks excavated at Halieis, one is certainly of this type (House D), but the other (Industrial Terrace), sunk in the ground under a cement floor, appears to be quite different, though both are from excavated contexts which are Classical in date (Ault 1994: 205–6, figs. 2–3). In the Southern Argolid survey publication (Runnels in Jameson *et al.* 1994: 272) Fig. 5.6 demonstrates how the wooden superstructure for the windlass might have worked with reference to Fig. 5.7, which depicts the weight block found at site E26 and describes it as 'Late Classical/Early Hellenistic' (Fig. 6.34). However, in the site catalogue, E26 is identified as a Late Roman farmstead (no Classical, Classical–Hellenistic, Hellenistic or Roman material was found). Moreover, the weight block was found with a press bed, the *orbis* of a *trapetum* and a possible stone roller (identified as fragment of a possible ancient column), all of which most likely go together as part of the same assemblage (Jameson *et al.* 1994: 489; cf. 385, table 6.6 where the weight block is also identified as Late Roman). The identification of this weight block as Classical or Classical–Hellenistic is not upheld by the archaeological context, which strongly supports a Late Roman date.

Fig. 6.34. Southern Argolid, limestone press weight block (Jameson *et al.* 1994: fig. 5.7, reproduced by permission of the Trustees of Stanford University).

Some of the pressing equipment from the Southern Argolid survey is undatable. Frequently this is because it is no longer in its primary context but has been reused for a function other than pressing (e.g. B6, E50, E81), but sometimes it is because it has been found on a multi-period site (e.g. B103, E70, G1), or because there was not sufficient datable material recorded on the site (e.g. B103, G14, D8). Viewed in terms of their archaeological settings, almost all of the press assemblages to which dates can be assigned with some probability would appear to be Roman or Late Roman (B78, E7, C11, E26, A8, B5, B91, E12, E45, F2, G12).[16] The few which seem most likely to be Classical–Hellenistic (E30, E38, A60, E52, A61) provide no clue as to whether they were used for processing olives, grapes, both, or something else entirely, except for A61, where the press bed was found near a basin still sunk in a cement floor and there were two 'pieces of columns' (possible rollers from a crusher?) nearby. This has the best claim to be an identifiable olive press of all those found by the survey (Table 6.2). The evidence contrasts sharply with the assertion that 'fifteen sites, used certainly or probably in this [Classical] period, have evidence of such [pressing] installations' (Jameson *et al.* 1994: 384).

[16] Jameson *et al.* 1994: 400 Table 6.9 also includes parts of *trapeta* from Ermioni (E19), Dhidhima, and Loutro (B20?). I have not included these since none are detailed in the site catalogue.

The paucity of presses which can be securely identified as Classical–Early Hellenistic and the lack of positive evidence that they were exclusively, or even primarily, used for processing olives undermines the Southern Argolid Survey's claim that the Late Classical–Early Hellenistic expansion in numbers of small 'farmstead' sites in the countryside is related to the production of olive oil as an income-generating crop (see also Acheson 1997). The group of small sites in the fertile Flamboura area, used as a case study to support the argument that the region displays an 'orientation to olive culture' in this period (Jameson *et al.* 1994: 386), amply demonstrates the problems. Here, seventeen of these small sites were 'mapped' as if each economic unit was constituted of a single discrete parcel of land farmed by a single household, and all sites were in use at the same time for the whole period (Jameson *et al.* 1994: 386–92; fig. 6.20, table 6.7). Such assumptions are almost certainly unrealistic since we know that in most of ancient Greece farmers generally owned land in fragmented parcels, not contiguous blocks (Foxhall 2002; Acheson 1997: 177–8). The problem is acknowledged by the Southern Argolid Survey, but 'circumvented' by suggesting that fragmented holdings are associated with residence in a nucleated settlement (Jameson *et al.* 1994: 390–1, especially n. 9). In fact, the presence or absence of fragmented holdings is less directly connected to residence patterns than to the inheritance system. If the kind of partible inheritance system documented from other areas of Greece in classical antiquity had been in use in the Southern Argolid, then land holdings were almost certainly fragmented, and the picture presented of discrete farmsteads, each with its own substantial parcel of land, as well as the postulated sizes of these holdings, is inaccurate.[17] Most importantly, only one of these seventeen farms (B78) has any evidence of pressing equipment, which in this case is probably later than the Classical period, since it includes fragments of a rotary crusher which must be Roman or Late Roman (Table 6.2).

[17] In partible inheritance systems the estate is divided equally among heirs. In ancient Athens, Gortyn, and elsewhere, it was normal for sons to divide agricultural land equally, encouraging land fragmentation, though collateral inheritance, inheritance through daughters in the absence of sons, and other social and political processes might also serve to recombine fragmented parcels of land. See Foxhall 1989; 2002; Burford 1993: 37–48.

6.5.4. The Atene Survey

The case for an economy based largely on olive cultivation is even weaker for the Atene Survey than for the Southern Argolid. Lohmann's survey of this unprepossessing area of southwestern Attica near the mining districts has revealed evidence of forty eight 'farmstead' sites of various types (Lohmann 1993: 136–84; summarized in Table 4, p. 137). There is evidence for only a few pressing installations. CH31, dubbed the 'Ölpressengehöft', has a press bed, undatable on its own but lying within a structure where only Classical pottery was found. The press bed may be Classical in date, but no other pressing equipment was found with it (Lohmann 1993: 374). Site LE16 has the central column of a *trapetum* bowl and a threshing floor near the ancient site. The *trapetum* fragment is unquestionably Roman–Late Roman and goes with the Roman–Late Roman pottery found on the site. This crusher probably goes with the press bed spout and press weight block found on LE17 about 200 m away as part of what was originally a single Roman–Late Roman pressing installation (Lohmann 1993: 513–16; 1994: 87–9 and pl. 2.2, 2.3). A fragment of the central column of a *trapetum* and a roughly cut press bed were found on site AN3 (Lohmann 1993: 495; pl. 69.2), again suggesting a Roman–Late Roman date for this pressing establishment. A counterweight stone for a press with two rectangular holes in the top (Brun's 'type 41') was found on site AN23 (Lohmann 1993: pl. 70.4). This could be Classical in date, but could also have been used later. A boulder with a square hole cut in the middle from PH2 is tentatively identified as part of an olive press (Lohmann 1993: 413 and pl. 101.4), which it almost certainly is not. This evidence is plainly insufficient on its own for postulating the large-scale cultivation of olives in the region during the Classical period.

The basis of Lohmann's argument that these small 'farmsteads' grew largely olives and grain is the discovery of decayed terrace systems around some of the 'farmstead' sites. These are indeed very interesting features of the landscape, but it is impossible to date them securely and positively. Lohmann enthusiastically declares them to be Classical on the grounds that 1) they are comparable with the 'classical' terraces which Bradford and others have located by aerial photography; and 2) there is no other period when the landscape was inhabited to which they could possibly belong. That they must be for olive cultivation, he argues, is because nothing else would satisfactorily grow on these slopes and growing olives conforms to modern practice.[18] However, as Acheson (1997: 175–6) demonstrates for the Southern Argolid, terraces have regularly been used for cultivating cereals as well as tree crops.

None of these arguments offers very strong positive evidence for considering the region a major olive-growing area in classical antiquity. The terrace walls and threshing floors on which Lohmann has relied for creating his picture of classical agriculture are in themselves undatable at our present state of archaeological knowledge: there is no datable artefactual material incorporated into them or deposited beneath them. Such walls and threshing floors could date from any time between antiquity and the present. Nor need the lack of evidence for nearby settlement preclude the use of this land for farming. It may have been cultivated by farmers resident some distance away, and historical and ethnographic examples of the exploitation of plots far from residences and settlements are well documented (Forbes 1982; 1993; forthcoming). I have argued in more detail elsewhere (Foxhall 1996: 60–1) that the dating of Bradford's 'classical' terraces should be questioned. They were never verified on the ground, and again could plausibly date from almost any period. Overall, the conclusions which Lohmann has reached concerning the interpretation of this survey material seem overly optimistic, especially in comparison with the evidence for Roman–Late Roman olive oil production from Methana and the Southern Argolid.

6.5.5. Data from Other Regional Surveys

It is clear that the low levels of ancient pressing equipment discovered by the Atene survey are not unusual. The Lakonia survey found only two items (Cavanagh *et al.* 1996: sites Q359 and J219), Kea only three

[18] Lohmann 1993: 194–219, especially 194 (the importance of the olive in Attica conforms with the presence of terraces for olive cultivation), 198 (no other crop is feasible and growing olives on such terraces is in keeping with modern practice), 202–3 (their antiquity is proven by other, similar terraces located by aerial photography and the lack of settlement in other periods indicates they could only have been built in classical times).

(Cherry *et al.* 1991: sites 9, 15, and 64), and Aetolia, discussed in section 6.5.1, only one (Table 6.2).

In Lakonia, a probable *orbis* from a *trapetum* was found on Q359, a find in keeping with the Roman and Late Roman pottery on the site. A rough press bed (?) cut in a boulder was found about 40 m NW of site J219, which seems to have mostly Classical and Hellenistic pottery and nothing later, though so little material was found, it is difficult to assign a Classical date to this installation with certainty.

On Kea, site 9 has the best claim to be a Classical press, though not necessarily primarily for processing olives. The location just above a small valley sounds ideal for vines, but would clearly suit many crops well, including olives. Virtually no pottery later than Hellenistic was found here (and no Roman), and there were many sherds of larger coarseware vessels, especially pithoi and amphorae. No potential crushing equipment was found. It is noteworthy that the terraces surrounding this site are certainly post–Classical: the survey team noted Classical pottery in the filling supported by the eroding terraces (Cherry *et al.* 1991: 78–9, 144 (Fig. 5.17). Site 15 seems to have been a multi-period 'farmstead' site, and the pair of press beds found here is likely (though not certain) to belong with the Roman–Late Roman component. There is no evidence of crushing equipment (Cherry *et al.* 1991: 83–4, 145 [Fig. 5.18], 296 [Fig. 13.9]). Site 64 features an interesting andesite millstone from a rotary crusher, with a relatively flat profile and striations along the outer edge. Since this multi-period site is now occupied by a modern farmhouse, and the millstone has been reused as a *vothros* cover, it is difficult to date it very precisely, though it is probably Roman–Late Roman (Cherry *et al.* 1991: 123 [no. 64.8], 156 [Fig. 5.34]). It may be significant that only one stone was found: to judge from the flattened profile and the narrow band of striations it is possible that it came from a single stone rotary mill similar to those at Maresha in Israel/Palestine, though the possibility that it was one of a pair from a *trapetum* cannot be discounted.[19]

[19] Kloner and Sagiv 1993: 121–2 and Figs. 1–2. There are Late Roman examples of striated millstones from *trapeta*, e.g. on Methana, MS109 (Fig. 6.26), but they are often smaller in diameter with a very rounded hemispherical profile.

6.5.6. Presses in the Greek Countryside

It is clear from this analysis that remarkably few examples of Classical–Early Hellenistic pressing equipment have been found by intensive survey of the Greek countryside. Even fewer of these can be identified with any certainty as dedicated olive presses. It is certainly possible that much Classical olive pressing equipment is archaeologically invisible because it was made of wood, and the evidence of the Attic Stelae (see below, section 6.6) could support this. The implication, that most 'presses' were small, 'modular', fairly simple, and often multi-purpose, also suggests that olive cultivation and processing was generally carried out on a relatively small scale, even by wealthy land owners.

The evidence for olive and wine processing from Roman and Late Roman Greece offers a striking contrast to the Classical period. The majority of pressing installations discovered in the Greek countryside date to these periods. Frequently they appear on the sites of classical 'farmsteads' reoccupied in the Roman–Late Roman period, sometimes after a considerable gap or gaps in occupation. In Methana and the Southern Argolid in particular the amount of pressing equipment approaches the densities of presses found in major oil-producing regions of the Roman Empire.[20] These data strongly suggest that for Greece, periods when oil and wine manufacture were widely pursued as income generating enterprises may be archaeologically distinguishable from those when only smaller-scale processing was the norm. It is likely that in the Classical and earlier Hellenistic periods, the large-scale, income-generating production of olive oil was not common. The olive was certainly not the revolutionary 'cash crop' grown for profit that it is sometimes made out to be (cf. Hanson 1999: 33–5, 78–84, who considers intensive arboriculture and viticulture a 'revolutionary' development of the eighth century BCE).

A useful analogy is offered by the very different distribution of grain milling equipment found in the course of intensive archaeological survey (Table 6.3). It is reasonably safe to assume that these implements, found outside urban centres, were intended in all periods primarily for

[20] For comparable data from elsewhere in the Roman empire see in general: Mattingly 1996a; Hitchner 1993; N. Africa: Mattingly 1988b; 1993; Mattingly and Hitchner 1993; N. Africa and Spain: Mattingly 1988c; Spain: González Blanco 1993; Israel/Palestine: Frankel *et al.*

Table 6.3. Grain mills and presses found in five Greek surveys.

SURVEY	Saddle Querns	H–R Gr. Mills	Rotary Querns	Total Gr. Mills	C–HE Presses	R–LR PA Presses	Total Presses	COMMENTS
Kea	7	4	1?	17	1	2	3	5 undatable millstones
S. Argolid	27 total: 13 def. 3 poss. C–HE; 9 def. 2 poss. R–LR	7+	7 definite 2 possible	46+	6	11	25	8 undatable press assemblages
Methana	1	4	1	6	3 definite 2 possible	25	47	
Lakonia	2	4	4 definite 1 possible	11	1?	1	2	
Atene				7	1?	2	3	Grain mills probably C–HE but description not sufficiently detailed to allow dating

household use, not for producing marketable commodities. Although a number of the comparatively 'high-tech' (and easily identifiable) hopper-rubber mills have been found, generally, the more basic saddle quern is the norm for the classical period. Although high-tech rotary querns were regularly in use during the Roman–Late Roman periods (Runnels 1990), they are not ubiquitous, and none of the larger, animal-driven mills are documented anywhere in the Greek countryside (towns, where there are specialist bakers, are a different matter). Overall, considerably higher numbers of millstones for grinding grain are found than pressing installations, their distribution is wider, and more are identifiably Classical–Hellenistic than Roman–Late Roman: of the four surveys tabulated in Table 6.3 only Methana contravenes these trends.[21] This underpins the conclusion that the archaeologically visible pressing equipment of the Roman–Late Roman periods is a manifestation of a very different ancient economy from that of Classical Greece.

6.6. PRESSING EQUIPMENT IN THE ATTIC STELAE

The Attic Stelae (*IG* 1³ 420–430; Amyx 1958; Pritchett 1956) are records inscribed on stone set up by the *Poletai* of the property confiscated from a number of wealthy Athenians towards the end of the fifth century BCE as the result of the trials following a major political and religious scandal (Thuc. 6.27–9, 60–1.). They document a wide range of agricultural tools and equipment, including some which could be for olive and wine pressing and processing. What it consisted of, how it was sold, and where it was stored, all suggest that presses and similar devices were multi-purpose machines, easily assembled from (often) multi-purpose components, and as easily disassembled for storage or use for another job. Why classical Greeks, or Athenians at

[21] The small number of grain grinding stones is not an artefact of faulty observation. Only one Classical–Hellenistic saddle quern was discovered and, with one exception, the fragments of rotary mills found all postdated the Late Roman period. Given that Methana is made of volcanic stone suitable for millstones, and that numerous saddle querns were found on prehistoric sites (e.g. MS108), this is very striking. However, many 'farmstead' sites featured ceramic and/or stone mortars (e.g. MS109), and these were perhaps used for grain grinding.

least, did not construct specialized, permanently installed presses will be reconsidered towards the end of this chapter (6.8.3).

One of the major difficulties of identifying such equipment in the Attic Stelae and other Greek sources is the impenetrability of the technical terminology. As we have already seen in the case of olive crushers, words for everyday objects can disguise quite specific pieces of machinery. For example, *τράπεζα* meaning a 'lower grinding stone' as well as a 'table', or the protean word, *ὄνος*, 'donkey', can mean 'a moving/working part' on almost anything from a quern to a windlass to a spindle. Often it is impossible to penetrate these terms fully.

Some of the assemblages sold off in the Attic Stelae are clearly the contents of country houses and their outbuildings. As assemblages, they reflect the mixed agronomic regimes of the agricultural enterprises of these wealthy Athenians discussed in Chapter 2. For example, the beginning of column 2 of *IG* I^3 422 contains just such an inventory (Table 6.4). The first readable (or at least restorable) entry is a clay kneading trough, followed by a short lacuna. This space may represent the shift in focus of the inventory from the kitchen area to another part of the establishment, since there is no indication of a change in location or owner at this point. The next seven entries are cloaks (*τρίβον*) of a type that poor people were expected to wear (Pritchett 1953; 208–9). These were probably clothing for slaves. These are followed by an illegible entry, then another lacuna, perhaps representing another shift, this time perhaps to an outbuilding or storeroom. The first two entries of this next section are unreadable, followed by four *medimnoi* of something; perhaps cereal of some sort is most likely? The next few entries are all food items: something unreadable, poor quality wine or vinegar, pickled olives. All of these items sound like food intended to feed slaves: low quality, cheap basic staples. The following entries contain a marvellous assortment of agricultural equipment: two winnowing fans, four threshing shovels/forks (see Benaki Museum 1978 for modern examples), (probably) two threshing sledges,[22] twenty eight roof timbers, one hundred-plus

[22] This interpretation of *ὀκίστια* is different from that of Pritchett 1956: 297–9, followed by Amouretti 1986: 107–8. As Amouretti herself realized, harrowing is nowhere mentioned as an important operation in arable cultivation in the Greek sources. This is because it was usually unnecessary because, at least on the holdings of wealthy farmers, arable land had received multiple ploughings before sowing. Black

Table 6.4 Assemblages of farm equipment from the Attic Stelae. (*IG* I³ 425; 422.col. 2).

<table>
<tr><th></th><th>Col. II</th></tr>
<tr><td>- - - -</td><td>[...⁵..μέ]διμνοι Δ</td></tr>
<tr><td>- - - -</td><td>[...⁶...ο]ν λέκος</td></tr>
<tr><td>- - - -</td><td>[ὄχ]σος [στ]άμνοι |||</td></tr>
<tr><td>[...⁶...]οκο ||</td><td>[ἐ]λαõν σ[τ]άμνοι ||||</td></tr>
<tr><td>[...⁶...]ς</td><td>[π]τέο ||· θρ[ί]νακε⟨ς⟩ ||||</td></tr>
<tr><td>[κάρδοπ]ος</td><td>ὀκίστια [Δ]||</td></tr>
<tr><td>[κεραμ]εία</td><td>στροτε̄ρε[ς] ΔΔΓ|||</td></tr>
<tr><td>lacuna</td><td>κεράμο στε[γ]α{σ}—</td></tr>
<tr><td>[τ]ρ[ίβον]</td><td>στε̄ρος ζεύ(γε) Η[...]|</td></tr>
<tr><td>[τ]ρίβ[ον]</td><td>γέρρα ||| γαλε[άγρ]α</td></tr>
<tr><td>τρίβο[ν]</td><td>σκαλίδες |||</td></tr>
<tr><td>τρίβον</td><td>τόρος· καρκίνο[ς?]</td></tr>
<tr><td>τρίβον</td><td>πέδε καὶ ἄχσον</td></tr>
<tr><td>τρίβ[ον]</td><td>δρέπανον</td></tr>
<tr><td>τρ[ίβον]</td><td>ἀμπελοργόν</td></tr>
<tr><td>π- - - - -</td><td>σμινύαι Δ</td></tr>
<tr><td>lacuna</td><td>δίκελλαι |||· τρυπ[άν(ο)] ||</td></tr>
<tr><td>- - - - - - - ς |||</td><td>ὀβελίσκοι Γ|- - -</td></tr>
<tr><td>- - - - - - - vacat</td><td>κρεάγρα ||</td></tr>
<tr><td></td><td></td></tr>
<tr><td></td><td>σύκον φορμοί |[||--]</td></tr>
<tr><td></td><td>λίτρο σαργάνα[ι--]</td></tr>
<tr><td></td><td>σεσάμο ἀπλύτο</td></tr>
<tr><td></td><td>ἐμισάκιον</td></tr>
<tr><td></td><td>χέρχνον ἐμισά[κιον]</td></tr>
<tr><td></td><td>μελινõν ἐμισάκ[ιον]</td></tr>
<tr><td></td><td>ἀμυγδαλõν φορμ[οί--]</td></tr>
<tr><td></td><td>κοριάννο φορμο[ί--]</td></tr>
<tr><td></td><td>κάδο πιττίνο[||]</td></tr>
<tr><td></td><td>τελίαι ΔΓ|||</td></tr>
<tr><td></td><td>σκιάδειον σ[μικρ(όν)]</td></tr>
<tr><td></td><td>βάθρα Γ· θρόν[οι--]</td></tr>
<tr><td></td><td>δίφρος·" νε---</td></tr>
<tr><td></td><td>[...]ο[...]"ν</td></tr>
<tr><td></td><td>[ἀ]σκέρα</td></tr>
<tr><td></td><td>[τ]ράπεζ[α]</td></tr>
<tr><td></td><td>[κ]λιντέ[ρ]</td></tr>
<tr><td></td><td>[δ]έσμε σ---</td></tr>
<tr><td></td><td>[.]|Δ---</td></tr>
<tr><td></td><td>lacuna</td></tr>
</table>

figure scenes of sowing sometimes show a man coming along behind the sower with a large mallet or mattock breaking up clods—all that was usually necessary (Fig. 6.42). Similarly, Xenophon (*Oec.*16.15) mentions preparing fallow by digging with mattocks, which would have again obviated the need for harrowing. Pritchett 1956: 298 recognized that the word ὀκίστιον is cognate with the Latin *occa* and Greek ὀξίνα. Hesychius describes an ὀξίνα as 'an agricultural implement having iron teeth, dragged by cattle'. If harrows were not regularly used, then ὀκίστιον is most likely to be a threshing sledge. That they follow other tools used for threshing in this passage of the Attic Stelae also suggests that 'threshing sledge' is the correct meaning. With the corrected translation it is perhaps more sensible to restore the number as two rather than as eleven, as in Amouretti 1986: 107 and *IG I*³, following Pritchett 1956: 297–8.

Fig. 6.35. Attic BF kylix (Paris Louvre F77) showing ploughing, sowing, and clod-breaking.

pairs of roof tiles.[23] The items which follow the roof could be connected with pressing: three flat basketry things (γέρρα) (Amyx 1959: 265–6), which are likely to be frails for holding grape or olive pulp on the press bed; and a weasel trap, which may be a genuine weasel trap but might also be framework to hold frails in place on the press bed.[24] These are followed by three hoes, which are general cultivation/digging implements.

The next few items are pieces of machinery, which could have functioned as parts of presses, but might also have a wide range of uses in other agricultural devices as well as for building and construction. *Τόρος* and *καρκίνο*[ς?] (the latter word literally means 'crab') both have a wide range of meanings as tools or parts of machines (Pritchett 1956: 303–4; 294). *Τόρος* means a piercing instrument, for anything from well-digging to stone-cutting; perhaps the latter meaning is more apt here. Pollux (10.148) may help on the *καρκίνος* for he says that in the Attic Stelae it referred to a device for lifting stones. Such a machine could, of course, also be used for lifting stones for use as crushers or press weights.[25] Since the last letter of

[23] These last two items may be the roof of the outbuilding itself.

[24] Hero mentions a γαληάγρα used for this purpose (Drachmann 1932: 60–2). His seems to have been a new, improved variety, but there is no reason that the word could not have previously carried this meaning.

[25] Compare the use of pulleys and winches for lifting weight stones for olive presses documented by Frankel and Avitsur in Palestine/Israel, Frankel *et al.* 1994: 41, 115, and figs. 35, 115.

καρκίνος is missing, and the word could therefore be in an oblique case, it is possible that the *τόρος* could be part of this stone lifting machine (i.e. the *τόρος* 'in' or 'of' the *καρκίνος*). *Πέδε καί ἄχσον* is another mysterious piece of machinery, with a considerable range of possible meanings and uses (Pritchett 1956: 289, 299). A windlass (perhaps with its brake) might be one of the more likely possibilities.

There follows a pruning hook for vines, ten mattocks, three two-pronged forks, two augers or borers of some kind, six or more spits, two meat hooks, several containers of dried figs, a container of sodium carbonate (perhaps for soap making, see Pritchett 1956: 311–12), half a sack of uncleaned sesame, half a sack of each of two different kinds of millet, containers of almonds and coriander seed, two large resin-coated jars (elsewhere associated with pressing, Ar. *Pax* 1202), and eighteen grain sieves. The 'small sunshade' which follows is not, I think, a parasol (Pritchett 1956: 209–10), but a small shelter of reeds or matting which can be taken out to the fields and set up to provide shelter for eating a meal or resting at midday. Most of the rest of the items are probably furniture: five stools(?), a chair or chairs, a backless chair, a pair of woolly shoes,[26] a table (or millstone?), a couch, and bundles or sheaves of something. The rest of the list is illegible. The diversity of this list makes the wide range of activities carried out very clear, including the cultivation of both arable crops and tree crops. In addition, many items strongly suggest the presence of slaves to do most of the routine work, for example, the substantial numbers of cloaks, mattocks, corn sieves, and stools.

IG 1^3 425.4–14 also lists the contents of an outbuilding or something similar, in which there are many items that can be identified as parts of presses (Table 6.4). Most obvious are the seven *πιεστερία*. This word can be used to mean the whole press (Amouretti 1986: 286), though it clearly does not mean that here: there are seven of them and they are not highlighted as large, expensive items. Amyx (1959: 251–2) was puzzled by these items and thought in desperation that they might clamps holding the broken bed together. More likely they represent some significant part of a press, the most likely candidates being: 1) press weights, 2) stone or wooden 'plates' to be placed on top of the heap of frails on the press bed, or even possibly 3) press beds.

26 If this is to be understood literally and does not disguise another piece of equipment.

The ρυμός, 'log, beam', which follows them must be the press beam itself. With this go the τροχιλεία, 'block and tackle', for pulling down the press beam and/or raising weights; and the hαμαλλεῖον, which as Pritchett (1956: 288–9) pointed out, is too expensive for its primary meaning of 'sheaf band', and perhaps ought to be translated 'hank of rope', to go with the block and tackle and the press beam. The resin-coated storage jar is also suggestive of pressing equipment, especially for wine. The other items are more difficult. The βάθρον could be a bench or stool, but it could be any number of other things as well. The word is often used to mean a base or stand so it could conceivably be a stand to take a press bed (Pritchett 1956: 215), like the one shown supporting the press on the Boston skyphos (Fig. 6.2). Alternatively, it could be some other unidentifiable part of a press. The 'broken pallet' may be a genuine bed, but like the English word 'pallet', this may be a technical term for some part of the press. The rotten double doors, seven rotten doors (in two lots), and the firewood, may have been placed with the other items just because this was a convenient storage area. But it could possibly indicate that this was an area where olives were pressed since fuel would be needed to heat the water used in the pressing process. The interesting feature about all the bits of press that we almost certainly see here is not only that they were disassembled, but that they were sold separately. These are not sophisticated and specialized permanent installations, but 'modular' machines, from which parts could be removed and used for something else.

The same is true of the wine treading area or press located in the Kerameikos (as reported to the *Poletai* by the demarch) which is sold in bits on column two of the same inscription (*IG* I^3 425.30–40) (Table 6.4). The two (?) λενοὶ λίθιναι are probably either large stone troughs (Brun 2004a: 91) or press beds, although the word ληνός can also be extended to mean the whole press (Amouretti 1986: 173–4; Amyx 1959: 242–6; cf. *IG* I^3 422.189). The two (?) clay [τρ]ὶπτε<ρ>ε could be the receptacles to take the must from the treading floor, or if there was a press here (which is not certain), the outflow from the press, if we follow the definition given in Pollux 10.130 (Amyx 1959: 247–9). However, this is another word with multiple meanings and it is also used for 'pestle' and sometimes 'mortar', which could serve a number of purposes including the crushing of olives in small quantities. The pithos mouth probably also had something to do with the channelling of liquid from the treading floor. The 'lead-mended mushroom of 1 chous capacity' must be a measuring jug, used in the treading/pressing operation (Amyx 1959: 208–11). The *φάτνη* could be a large vat as well as a chest or table (Pritchett 1956: 243–4). And, the [πλ]*ίνθοι σταφυλοβόλοι* are most likely to be the tile or brick or stone surface of the treading floor itself (Amyx 1959: 249–50). I am totally baffled by the *βαθρόθυμα*. A 'stand for an incense burner' sounds completely wrong here, but I have no idea what it might be.[27] Again, the whole establishment is clearly for one operation, but the bits are not specialized, highly crafted parts, and are considered to be separable for purposes of sale, and possibly use.

These examples from the Attic Stelae make clear why the archaeological remains of Classical Greek presses look so pathetic compared to their larger and more specialized Roman relatives (Drachmann 1932, Amouretti and Brun 1993, White 1984, Mattingly 1988a; 1988b; 1988c, 1996a; 1996b, Mattingly and Hitchner 1993; Brun 2004a; b). The epigraphical evidence suggests that pressing establishments were small, and not necessarily permanently installed. Because the component parts were devices that could be used for other purposes, they probably often were, only to be reassembled when the press was needed again. Nor did every farm unit, even those belonging to wealthy landowners, have a press associated with it: very few can even be guessed at in the Attic Stelae. The probable reasons for this will be discussed in section 6.8.

It is also significant that there is no trace of anything in these texts which looks like a rotary olive crusher or part of one. If such machines were going to be visible in the documentary record, one might expect to find them here. They emanate from a still-imperial Athens, and testify to the assets of wealthy landowners in Attica and abroad. It is clear from these documents that these farmers grew olives and produced olive oil, and that they owned pressing equipment. Indeed, a substantial wine-producing estate documented in these documents was discussed in Chapter 3.6. Rotary crushers would have been expensive pieces of equipment worth selling (in their entirety or in bits), and they would not have been easily hidden from the *poletai*. But, that even these

[27] Pritchett 1956: 309. Unless incense discourages the wasps which congregate when grapes are being trodden? One man on Methana used to tread his grapes in rubber boots so as not to get stung.

prosperous land owners do not seem to have used rotary crushers indicates that they were probably not in use in Attica at least at the end of the fifth century BCE, and that the mid—fourth century BCE date for their introduction suggested by the archaeological finds of these olive crushers is correct.

6.7. THE DELIAN COUNTRYSIDE

The countryside of Hellenistic Delos is documented epigraphically and archaeologically in a rather different way from that of Attica. On Delos and the neighbouring islands of Rheneia and Mykonos, estates owned by the Sanctuary of Apollo were leased out, generally for five years at a time, and the details of these contracts were recorded by temple officials on stone (Kent 1948; Brunet 1990a; Reger 1994). The documentation is particularly full for the period between about 314 and 166 BCE. These properties were leased out to wealthy land owners who appear to have run them as income-generating 'business ventures'. Crops, trees, buildings, and equipment appear in the inventories of these estates, but the general principle seems to have been that movables (including wooden doors and windows in some cases) belonged to the lessee rather than to the Sanctuary. It is interesting that, except for two estates on Mykonos, these farms appear to have no olive trees, although many vines are recorded. Neither are there any obvious presses or treading floors listed, although several farms have a 'mill' (*mylôn*), whatever that might mean. These were not necessarily grain mills, since most grain grinding in this period was done with saddle querns or hopper-rubber mills, and both types are small and portable enough that they seem unlikely to be equipment owned by the landlord (i.e. the Sanctuary). Despite finds of presses in the urban site of Delos dating to the later Hellenistic period (second—first centuries BCE), archaeological finds of pressing equipment appear to be largely absent from the Delian countryside, although there has been no systematic intensive survey of the island. One exception is a granite wine treading floor, thought to be located on a farm mentioned in the inscriptions as having numerous vines (Brunet 1990a; Brun 2004a: 92), although interestingly the treading floor is not listed as part of the farm equipment in the epigraphical record. The paucity of permanent, specialized, equipment for processing olives and grapes in the archaeological and epigraphical record may well suggest that such equipment was generally on a relatively small scale, and was adaptable and impermanent, as seems to have been the case elsewhere in Classical and Hellenistic Greece.

In the past it has been suggested that olives were unimportant or even non-existent in the Delian economy, but more recently Brun and Brunet (1997: 605–9) have convincingly argued that because these estates were subject to short, five year leases for much of their history, this provided a disincentive to farmers to plant olives, which are productive only in the long term. These leases may therefore not reflect the situation on privately-owned land in Delos and Rheneia (Reger 1994: 137–8). The situation on Sanctuary-owned farms may have changed during the second century BCE when the Athenians reoccupied Delos and the tenure of many leases increased to ten years (Brun and Brunet 1997: 608–9; Brun 2004a: 109). Reger (1994: 201–6) has observed that even though none of the leased farms appears to have any olives, levels of rents are closely correlated with oil prices (as recorded in Sanctuary accounts) in the period after 270 BCE. One reason for this may be that in years when the price of oil was high, the Sanctuary needed larger revenues for its purchase, which it raised on the rents of the farms. It is perfectly possible that the Sanctuary was buying a considerable amount of its oil from the very people who rented the Sanctuary farms, as they may well have grown olives on the land they owned privately. For the Sanctuary officials, this would provide a neat way of thwarting local wealthy landowners tempted to exploit high oil prices to their own advantage: as the price at which they sold their oil went up, so did their rents.

6.8. PRODUCTIVITY, PRODUCTION, AND DOMESTIC CONSUMPTION OF OLIVE OIL IN CLASSICAL GREECE

6.8.1. The Productivity of Olive Trees in Oil

It has already been noted that the returns from olive trees are highly unpredictable from harvest to harvest. This makes it very difficult to extrapolate any valid figures for 'average' productivity even just working with modern data, let alone attempting to apply such figures to classical antiquity. In trying to apply modern production statistics to the ancient world, one is hampered by the differences in cultivation techniques between then and now. In modern times the cultivars used may well be more productive of oil than those of classical antiquity. Moreover, the use of chemical fertilizers also has greatly increased yields. But these factors might be balanced against the possibility that much more intensive cultivation strategies were used in classical antiquity, with high inputs of labour and sometimes of organic fertilizers compared with many modern agronomic regimes. On the land of wealthy proprietors who owned slaves, this might have made yields as high or almost as high as they are now, but of course this is impossible to quantify. In addition, the rich landowners of antiquity might well have grown at least some of their olives on much richer soils than those on which olives are now most often grown. So, on the problem of 'average' productivity, I have assumed that the modern statistics I have available are not radically different from ancient productivity on the estates of wealthy landowners, though productivity on small-scale farms might well have been considerably lower. As this assumption is not testable and may be incorrect, the following represents only a best guess.

The olive oil production statistics from Kranidhi, Southern Argolid, Greece which cover the twenty year period from 1960 to 1980 vividly show the variability in yield (Fig. 6.36). These figures were collected from oil pressing establishments for statistical purposes, not for taxation, but they are nonetheless likely to be under-reported since they came from their recorded accounts. They suggest an average yield of around 2.6 kg oil per tree per olive harvest (*elaiona*) for the decade 1961–70, and 3.4 kg oil per tree per *elaiona* for the decade 1971 to 1980. The increased productivity may represent the larger number of hydraulic, diesel-powered presses in operation in the later decade. Ghiannakaris' (1985: 14) statistics on olive production from Khalkis over a much shorter period (1976–1979) give an average annual return of around 780 kg fruit per ha = 1560 kg fruit per *elaiona*. At a fruit:oil ratio of 4–5:1, and allowing around 80–100 trees per ha this puts average oil production per *elaiona* at between 3.12 and 4.8 kg oil per tree, very comparable with the Kranidhi statistics.

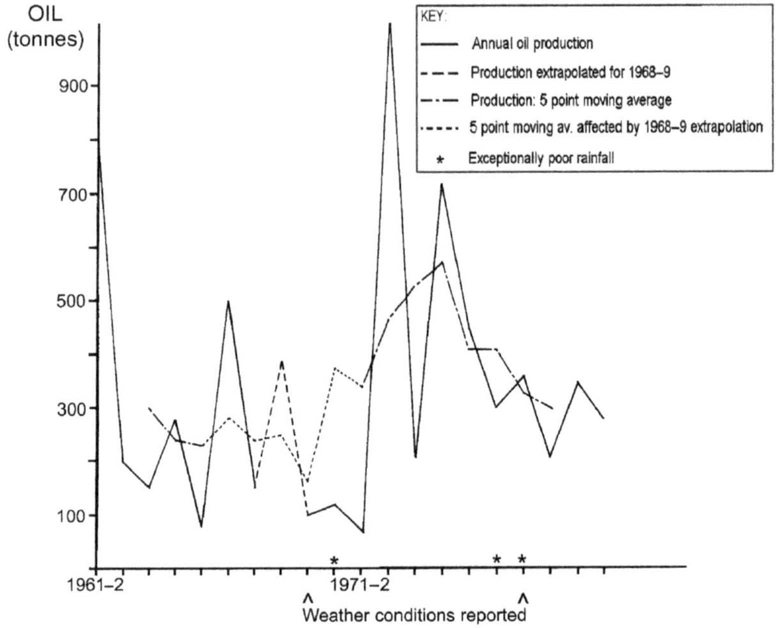

Fig. 6.36. Olive oil production in Kranidhi, Southern Argolid, Greece, 1960–1980 (H.A. Forbes).

6.8.2. The Efficiency of Ancient Olive Presses

The efficiency and through-put of Roman oil presses has been examined by Mattingly and Hitchner (Mattingly1988a; 1993; Mattingly and Hitchner 1993). Mattingly (1988a: 182–3; 1993: 484) has pointed out that in olive pressing there is effectively a trade-off between the time taken to press a load of olives and the pressure applied. So, improved olive presses, even modern hydraulic ones, do not greatly increase the absolute amount of oil than can be extracted, but they do considerably reduce the amount of labour, and even more the amount of time, taken to extract it. Sordinas (1971: 32) reached similar conclusions on traditional pressing techniques on Kerkyra (see also Frankel *et al.* 1994: 124). According to his data, it took two to three men four to five hours to do two pressings of a load of pulp on traditional screw presses, while it takes two to three men twenty minutes to press twice as much pulp on a hydraulic press. The hydraulic press increases the yield of oil by around eighteen per cent, but the difference might well have been less had three pressings been done on the traditional screw presses. That only two pressings were done presumably indicates that labour was in short supply.

The implication for the simpler and smaller Classical Greek lever presses is that although rather less oil would have been produced than on modern hydraulic presses or a large Roman screw press, given high inputs of time and labour, they would still have yielded at a reasonably favourable fruit:oil ratio. Mattingly (1988a: 182, 184–5) used a ratio of 4–5:1 of fruit:oil for Roman presses, based on a combination of literary and archaeological evidence. This is the same as the figure used for calculating fruit:oil ratios for the modern Greek data. A fruit:oil ratio of around 4:1 for classical Greek presses is therefore reasonable, but such a return would have taken a very high input of time and labour, significantly higher than on Roman presses, which were capable of exerting greater force. Obviously, if the time spent pressing were decreased, the oil yield would be lower. Presumably, presses were set up to press pulp almost 24 hours per day during the olive harvest, though during much of that time they would have needed little attention.

This suggests that the number of trees which could be served by a single olive press was lower in classical Greece than in Roman Italy or North Africa, or eighteenth–nineteenth century Greece. Venetian statistics from Southern Argolid in the eighteenth century suggest a remarkably consistent ratio of around 400–600 trees per press (Forbes 1993). The ratio of trees to presses was likely to have been, if anything, lower in classical Greece, since presses were quite small and incapable of exerting very high pressures (as shown above, section 6.5, 6.6.2). By contrast, if Cato's olive grove were serviced by his five presses this would give a ratio of around 900–1200 trees per press, probably toward the lower end of this range.[28]

[28] Avitsur in Frankel *et al.* 1994: 124, claims that one 'traditional' (normally screw) press can serve around 3300–7500 trees, based on an average crop of 10–15 kg fruit per tree, with the press installations processing about 80–100 kg olives per hour (about 2000 kg per day in peak season), though not all traditional presses operated at such high capacities. Such presses therefore were extracting a maximum of 2–3 kg oil per tree.

6.8.3. Conclusions: Numbers of Trees, Presses, and the Domestic Consumption of Olive Oil

The domestic consumption of olive oil for a wealthy household in classical Greece has been estimated at around 200–330 kg per year (see above, Chapter 4). This would be around 400–660 kg per *elaiona*, assuming that there was one major olive harvest every two years. At an average productivity as low as 2.16 kg oil per tree per *elaiona*, 185–306 trees would fulfil this requirement. This number is little more than the number of trees well-off small-scale farming families in Methana owned in the 1970s and 1980s. This would be around 1.9–3.8 ha olive trees if they were evenly spaced at around 80–100 trees per ha (which they almost certainly were not in most cases). Obviously, if oil yields per tree were postulated to be higher then even fewer trees would be needed. Some households, of course, had many more trees than this, but it is clear from the tone of Demosthenes 43 that Hagnias' 1000 trees constituted a sizeable enough orchard to be noteworthy (see above, Chapter 5.6), and, as discussed in Chapter 5.9, there were considerable numbers of sacred olive trees (*moriai*) in Attica. However, such calculations are highly speculative and must only be understood as roughly setting out the parameters of the possible. They are certainly not intended to be read as fact.

It is also clear from the number of trees per press in the Venetian and Roman examples quoted above that even if Classical Greek lever presses took much longer to extract oil than their later, more efficient counterparts, it was not worthwhile for households whose holdings of olives were at the lower end of the suggested 185–306 trees to have their own press. They might better pay a neighbour for the service, in cash, or in kind, or in some other way. For households owning 200 or more trees who had their own press, even if they took in other farmers' olives as well, it did not matter if the presses were slow. And finally, if presses were slow, there was no need to have high-powered crushers, producing pulp faster than the presses could process it. The olives might better wait for processing whole. This presumably explains why few farmers seem to have adopted the rotary crusher in classical times. Furthermore, the bottlenecks caused by the long periods when the press just had to be left sitting may have freed up some workers to go out and pick more olives.

Processing at this scale probably functioned perfectly adequately most of the time, though bumper years might have been exciting and the wastage rate at such times could have been significant. The whole system depended on very large amounts of labour to operate, but needed much lower investments in highly crafted equipment. With even a relatively small slave workforce this was almost certainly assured. Hence, even the wealthy landowners whose possessions were listed as sold on the Attic Stelae had no need for large-scale presses and crushers to supply their needs in olive oil. In this context, it is understandable why relatively low numbers of olive pressing installations have been recovered archaeologically via both survey and excavation, and why there appears to be a virtual absence of Classical pressing installations dedicated to olive processing documented in the archaeological record.

7

Arboriculture and Ornamental Gardens in Ancient Greece

7.1. INTRODUCTION

It seems very odd that the Greeks are not better known for their ornamental horticulture in all its many forms; indeed it has been suggested that the Greeks never made gardens for pleasure (e.g. Burford 1993: 137; cf. Osborne 1992b; Carroll-Spillecke 1989). On the contrary, there is a considerable of documentary, archaeological, and iconographic evidence to suggest that they did. Greek gardens were not, of course, herbaceous borders behind the house designed in the style of temperate-zone English or American gardens. It is hardly surprising, given ecological parameters of southern Greece, that they share much more in common with the gardens of southern California. Moreover, the word 'garden' itself is misleading: most scholars have tried to translate that back into ancient Greek as *kepos*, which demonstrably does not mean ornamental garden most of the time (Osborne 1992b).

The Greek vision of the landscape in which they lived is more complex than the dichotomies such as 'town and country' or 'wild and tame' in which they often expressed spatial relationships and meanings in literature and philosophy. On close inspection it is really a spectrum ranging from the urban space of the town to the wildest mountainsides. Both extremes, along with the many different kinds of spaces in between, are conceptually part of the polis and each has its own activities and uses associated with it. Different values were attached to each sector, though these values are not always fixed and may change over time and in special circumstances. Nonetheless, it is probably fair to say that most of the time, the 'most civilized' sections of the landscape—the town and the cultivated fields—had the highest value placed upon them in the sense of both economics and esteem. This is important for discerning what Greeks thought was ornamental in horticulture.

Much of what Greeks thought was beautiful in the design of outdoor environments we would probably consider boring today. Literary sources in combination with visual evidence suggest that they seem to have felt that the most aesthetically appealing landscapes were those which were most closely controlled and ordered. And, on a scale of wild to tame, places with ornamental plantings most clearly and powerfully manifested the subjugation of the natural world to human culture. This subjugation of nature to culture is central to Greek notions of the ornamental. Demeter was worshipped as Thesmophoros, the bearer of civilization, because she brought agriculture to man. Demeter is of course most closely associated with cereal cultivation. But, Kore, who comes along with her, brings flowers, and is regularly represented holding a flower (Fig. 7.1).

Order and symmetry were integral to the representation of the triumph of human control over the landscape, so perhaps it is not surprising that they become essential elements of the principles of beauty in general and decorative planting in particular. Regular patterns and lines, composed of well-shaped and balanced forms of individual plants, constituted the most pleasing kind of planting. So, for example,

Fig. 7.1. Attic amphora showing Triptolemos (centre) with Kore (left, holding flower) and Demeter (right) (Providence RISD 25.083).

when Xenophon (*Oec.* 4.21) portrays Lysander looking over the Persian gardens of Cyrus at Sardis, he admires the beautiful trees, the straight rows, the accurate angles, and the precise measurement.

Furthermore, they say that when Lysander came to him with the gifts from the allies, this Kyros was very friendly towards him, as Lysander himself once related to a friend in Megara, adding in particular that Kyros has personally shown him the *paradeisos* at Sardis. When Lysander had expressed amazement at the beauty of the trees in it (for they were planted at equal intervals in straight rows and all at regular angles), and many sweet fragrances wafted about them as they strolled around, he exclaimed in amazement, 'Kyros, I am certainly astonished at all these things for their beauty, but I admire even more the man who measured out each of the trees for you and arranged each one of them in order' (Xen. *Oec.* 4.20–1).

Where in the landscape of the Greek city, between urban centre and mountainside should we expect to find plants used decoratively? Ornamental planting was not limited simply to bounded 'gardens'. Certainly *kepoi* could be ornamental—many of those owned by the wealthy clearly were, but not all. Osborne (1992b) has pointed out that one of the main defining features of a *kepos* is the intensity of its cultivation. Whether it was ornamental or not was irrelevant to the designation of a plot as a *kepos.* The implication of this is that we need to look beyond the *kepos* to see plants grown for beauty and pleasure. Ornamental horticulture, therefore, appears as a feature in many different niches within the 'tamed' part of the landscape—within and adjacent to city and country houses, in towns, along roadsides, in villages, and even amongst the less remote fields. It is not a feature of the 'wilder' end of the spectrum of the human space of the polis.

Similarly, whether or not a plant had an economic value appears to have been irrelevant to its status as 'ornamental'. What made it ornamental was the way in which it was grown. One of the most striking features of Greek ornamental gardening is that many of the same cultigens and cultivars were grown for both for pleasure and for use. Useful produce from a decoratively planted plot was not perceived as contradictory. Nor were these gardens in the sense of flowerbeds; most Greek decorative horticulture on larger plots consisted primarily and fundamentally of trees, with other kinds of plants subsidiary to them. The emphasis on trees as the most important kinds of plants is evident in our sources. For example, Theophrastos, in his botanical works, devotes most of his books to trees, and trees are always discussed before other kinds of plants. He is particularly interested in distinguishing and comparing wild and cultivated trees (*HP* 3.2). He also explores a wide range of techniques for growing domesticated trees, including exotic imports (cf. Xen. *Oec.* 4.14: the Persian *paradeisos* is full of 'trees and all the other nice things the earth grows').

Although this might at first sound peculiar, the emphasis on domestic trees in ornamental horticulture is readily understood in terms of the notion that the most beautiful piece of land consisted of that which was most obviously under human control. The 'wildest' parts of the landscape in most of Greece consist of woodland, ranging from scrubby pine forest to mixed deciduous species. Even more frequently the uncultivated landscape consists of dense maquis—a tangle of unfriendly, shrubby vegetation full of large spines and biting seeds which is the bane of archaeological surveyors and shepherds alike (Grove and Rackham 2001). The terrain of these areas is often rugged and broken steep slopes, and the plants these areas support are tough and able to withstand the hot, dry conditions. In contrast it is easy to see the appeal of a flat or gently sloping plot with the odd well-placed rock, systematically planted with tall, well-grown and regularly shaped leafy trees as its main feature. It is also readily apparent how such a contrast might be made to symbolize the beauty of complete human control over the natural landscape.

Ornamental planting, then, occurred in large and small spaces across the domesticated part of a city's territory both inside the town itself and beyond it. Different plants and plant combinations were clearly appropriate for different kinds of spaces. I shall start by examining larger areas of ornamental planting, and move on in decreasing order of size and scale.

7.2. ORNAMENTAL ORCHARDS

For wealthy Greeks, the ornamental orchard was for long the primary form of ornamental horticulture. We know a surprising number of them from the sources—even Homer describes one laid out on virtually the same principles as later classical ones (Alkinoous' *orchatos*: Odyssey 7.112–131; cf. Laertes orchard: 24.221–7, 341–4). I would argue that from at least the later sixth century BCE (and perhaps earlier) most wealthy landowners aimed to plant at least some of their land near roads and settlements as ornamental orchard.

Combining the information provided in both documentary and iconographic sources it is possible to visualize the ornamental orchards of classical Greece quite clearly. Normally they do not seem to have been fenced, but a line of large trees grew along the outer edges, generally as part of the overall planting pattern. This would create an obvious boundary, though in most cases it would not be impenetrable. Nor would the plot have been invisible—being seen was crucial to their prestige value. Several species are mentioned as perimeter trees including plane trees (*Plantanus orientalis*) and poplars (*Populus nigra, Populus alba*), but olive trees seem to have been most regularly used. They are long-lived, grow very large, and provide good shade as well as a prominent borderline. And situated on the edges they do not overwhelm smaller tree species as they might if grown within an ornamental orchard.

7.3. ORNAMENTAL ORCHARDS: WRITTEN SOURCES

The literary sources regularly refer to planting olives along roadsides (Xen. *Oec.* 19.13; Dem 55.22). However, in the context of an ornamental orchard this line of olives became part of the planting pattern. Within the perimeter of olives or other large trees, a formal planting pattern of concentric circles appears to have been used, with the large trees bordering the road (see Chapter 5.8). In Aristophanes' *Acharnians* 991–9, the old Acharnian men of the chorus wax lyrical about the joys of peacetime, which include planting an ornamental orchard:

> Would that some Eros could join up you and me
> with a crown of flowers, just like in the picture,
> Or perhaps you've thought me rather elderly?
> But taking you I think there would still be three things to add,
> First would be to set out a long trench of little vines,

> Then alongside this new little rooted cuttings of baby figs,
> And the third trench of domesticated trees, since *I'm* an *old* man,
> Is olives all around the field in a circle, so that I can smear you
> all over with oil (and me) on the first of every month.
>
> (Ar. *Ach.* 991–9.)

A similar planting pattern for a very up-market ornamental orchard including flowers for garlands and wreaths and fruit trees is described by Apollodoros ([Dem.] 53.15–6) when he tells of his quarrels with his next-door neighbour Nikostratos:

> 'Coming to the field by night, he hacked about as many trees of table fruits as were planted there, and the vines climbing up them, and he broke down the young olive trees planted in a line around the borders of the plot so that not even invading enemies would have done so much damage. In addition, they sent in during the day a citizen boy, since they were neighbours bordering the field, and they ordered him to pluck the budding rose bush...' ([Dem.] 53.15–6.)

The formal nature of this decorative orchard is made clear by Apollodoros' tale of woe. The planting pattern consists of fruit trees—perhaps several different kinds—with vines climbing up at least some of them and rose bushes interplanted amongst them, most likely in a formal pattern. Again, a line of young olives has been set round the perimeter of the plot. The fruit trees were for table fruits, grown as delicacies. They almost certainly included some figs, which are frequently mentioned as grown with vines climbing up them (Theophr. *CP* 3.10.8; 5.5.4). But they quite likely included almonds, apples, pears, pomegranates, and plums as well. The rose bush is specifically designated as 'budding/in bud'. This may well be to distinguish it from other rose bushes planted on the plot which were not yet in bud (and hence would not be so damaged by having the shoot ends broken off).

This plot sounds rather attractive. The mix of larger and smaller trees with shrubby roses planted underneath and climbing vines would give an attractive range of heights and welcome summer shade. There was probably an interesting mix of foliage colour and texture, as well as a pleasing succession of flowers across the season with fruit blossom beginning early in the spring, followed by roses, then summer and autumn fruits, including grapes and brightly coloured rose hips.

The fruit trees are specifically denoted as 'grafted'—indeed the descriptions of the plants and trees are quasi-technical—i.e. they are in the gardeners' language of a good seed catalogue or gardening magazine, but they are not as technical as the *Flora Europaea* or even *The Grafter's Handbook.* This is not accidental. After all, it was in Apollodoros' immediate interests to emphasize the beauty and great value of his damaged property. Moreover, the possession of such an orchard is in keeping with the place he feels he deserves to hold in Athenian society.

This passage also highlights the fact that there is no special Greek word for what I have called an ornamental orchard. Apollodoros called his a *chorion*—simply a 'plot' or a 'field'. So does Aristophanes. So, too, do most of the other Athenian sources. Interestingly, in contrast, Xenophon uses the word *kepoi* for Persian *paradeisoi* (*Oec.* 4.13). The key here is that the ornamental orchard is not a special kind of place—it is a special way of planting. *Choria* can be located in town or countryside. Orchards could be planted more and less ornamentally—this was a spectrum, just as despite the reputation of 'English gardens', not all are equally pretty or full of plants. Nonetheless, landowners who could afford to do so clearly made an effort to plant at least some of their plots of trees attractively. Perhaps the nondescript, non-specific term *chorion* used in democratic Athens at least was intended to blur the eliteness of these plots and their owners. Certainly, the contents of Apollodoros' plot will have produced the kinds of 'semi-luxury' commodities which confirmed the eliteness of the consumers (flowers, choice fruits, fine wine, and oil).

The use of garlands and wreaths offers a good example of the way in which the produce of ornamental orchards was consumed. Crowns of leaves and flowers were normal garb on any occasion which was 'special'. In addition, diners at symposia are shown hold and smelling scented flowers presumably given out by the host (usually they look like lilies). Athenaios devotes a large chunk of his fifteenth book to garlands and wreaths, and provides a potpourri of Classical and Hellenistic quotations about them. Clearly, garland-making at its height was an art akin to modern flower arranging—the right plants with the appropriate significance had to be matched to the occasion. Flowers were chosen for both their beauty and their perfume, while aromatic foliage plants like myrtle were felt to relieve the effects of alcohol. Though there must have always been those who simply did the equivalent of flinging the flowers into a jug, the best garlands were a total experience of interwoven colours, textures and aromas crafted in flowers and foliage.

There are numerous other mentions of orchards which were ornamental to a greater or lesser extent. Xenophon's Ischomachos assumes that most of his peers will aim to plant trees in regular orderly patterns (*Oec.* 20.3). In Demosthenes 55.13, the plot with olive trees along the road edge also includes vines, figs, and family tombs. Again, it is possible that they were planted in a regular pattern—the presence of tombs may have encouraged ornamental planting, since there are references to planting white flowers on tombs (Nikander in Athenaios 15.684d) and using white flowers at funerals (Theophr. *HP* 6.8.3). And in two examples of crops in the field sold in the Attic Stelae (*IG* I^3 422.81–6, 87–9) the produce is listed as 'figs, grapes and olives'. At least one of these plots is urban and the small amounts listed suggest a small plot. This again may imply some degree of ornamental and/or formal planting arrangement.

From Theophrastos' later fourth-century descriptions of plants and their uses it is clear that many other plants were selected and fostered in such orchards for their ornamental value. Other climbers besides vines were trained up trees—ivy (used for garlands) was one of the most common. Indeed, Theophrastos (*HP* 3.18.8) makes clear that several varieties of variegated ivy were known, apparently including both gold and silver leafed varieties. These would have been extremely attractive climbing deciduous trees in winter, providing year-round interest and colour.

Plants like roses were carefully selected for their qualities of scent, flower size and complexity and even repeat flowering—Theophrastos (CP 1.13.11, 12) provides the earliest reference I have found to repeat flowering roses and he knows a number of variations on the basic *Rosa gallica* and *Rosa centifolia.* Fruit trees were bred to provide a range of early and later flowering and fruiting varieties (Theophr. *CP* 1.18.3; 2.11.6 'early apples'), and seedless (or perhaps less seedy) varieties of table grapes and other fruits were developed (Theophr. *CP* 1.21.2). Even fruit varieties with particularly ornamental blossoms were developed (*HP* 1.13.1—deep pink almond blossom).

Grafting as a significant element of plant propagation was well developed for both ornamental and economic purposes. Theophrastos (*CP* 1.6.1–10; 2.14.4–5) clearly distinguishes between the three main types of grafting (the general word is *emballô, embolê*): bud grafting (*enôphthalmizô*), whip grafting (*emphyteuô*), and crown grafting (*engkentrizô*) (Fig. 5.1). He is aware that plants can be grafted onto rootstocks of a different species and that several varieties of scion can be grafted on to the same stock (*HP* 2.1.4; *CP* 1.6.10). And he is familiar with the wide range of plants, trees and shrubs which can be improved by grafting, or indeed, which are only successful using grafted specimens since they do not grow true to type from seed.

The techniques of ornamental gardening were also well developed by the fourth century BCE. Though the elegant topiary of Roman gardens was not a feature of Greek ornamental horticulture, restricted growth forms were utilized for their ornamental value. Training plants up trees has already been mentioned. But climbers were also planted on stakes (Theophr. *HP* 2.1.2) and plants trained up walls or cascading over them (thyme: Theophr. *HP* 6.7.5). Pruning was clearly quite well developed and its effects on shrub and tree growth were clearly well known. Shrubby trees such as myrtle (*Myrtus communis*) and bay (*Laurus nobilis*) could be pruned and trained in several different ways, depending on the location and the ornamental effect desired. Much pruning may have been done informally as part of cutting foliage and flowers for use. Techniques such as festoon training are also mentioned by Theophrastos (Theophr. *CP* 2.9.4—festoon-trained pomegranates), and he is well-acquainted with its effect of encouraging flowering lateral growth. Many of these orchards seem to have been at least partially irrigated (Theophr. *CP* 1.19.5)—and the effects of added water and lack of water were also well known to Theophrastos.

Theophrastos (*HP* 2.2.10) was clearly very interested in planting exotic trees in his orchards. He was particularly obsessed with the date palm, which he never could persuade to fruit in Greece, however hard he tried (a bit like ornamental olives in Britain), and large sections of his botanical works are devoted to its idiosyncrasies (Theophr. *HP* 2.4; 2.6.1–12; 2.8.4; *CP* 1.2.1, 1.9.2, 1.20.2, 2.3.3, 2.3.7, 2.5.3, 2.9.3, 2.9.15, 3.17.1–8, 3.18.1, and this list is certainly not comprehensive). Various primitive forms of citrus (probably citron) were also grown as exotic ornamentals in orchards (Theophr. *CP* 1.11.1); perhaps even before anyone had devised uses for their fruits, or perhaps used as ingredients for perfume and perfumed oil. The use of exotics may appeal for reasons beyond their novelty value. For imperial powers, or those with imperial pretensions (or pasts), appropriating other peoples' plants can be a symbol of domination over their culture.

7.4. ORNAMENTAL ORCHARDS: ICONOGRAPHIC SOURCES

There is more evidence for ornamental orchards than simply literary and epigraphical sources. Several Attic vase painters depict ornamental plantings with considerable care—though the trouble some have taken to evoke an orchard setting is not necessarily obvious unless you have some idea what a rich man's ornamental orchard might have looked like.

Lucilla Burn has described the landscapes of the Meidias Painter, active in the late fifth century BCE, as 'paradise gardens' (Burn 1987:19). Fantasy they certainly are, but based on the reality of contemporary Attic private and sanctuary ornamental orchards and gardens. Recognizing the reality underpinning these settings both enhances their interpretation and reveals the true extent of their innovation.

In the Meidias Painter's name vase, a hydria in the British Museum (E 224; Burn M5), two scenes in ornamental orchards are depicted. On the shoulder, the rape of the Leukippides by Kastor and Pollux occurs in a sanctuary complete with cult statue and altar. The Dioskouroi swoop down on the unsuspecting girls from the upper part of the shoulder. There are three olive trees—one by each of the handles and the upper part of an olive tree in front of the altar near the centre of the picture (the right hand bottom side of the altar is also out of the frame of the picture) (Fig. 7.2). These are probably intended to signify the perimeter boundary of olive trees around the boundaries of the sanctuary 'orchard'.

Fig. 7.2. The rape of the Leukippides in a sanctuary garden (upper register) and the garden of the Hesperides (lower register), on a vase by the Meidias Painter (London, British Museum E224) (Burn 1987: pl. 3 [M5]).

The tree in front of the altar is particularly interesting. The painter has presented a scene which the viewer sees from outside the sanctuary, in which the positions of the olive trees, especially that central one, must represent an attempt to use the spherical body of the vase to augment the illusion of viewing a plot bounded in olives from the foreground. The sanctuary itself appears to have cultivated flowers growing in it—the flower picker is gathering identifiable lilies.

In the lower scene Herakles' trip to the golden apples of the Hesperides is shown. Here again, the ornamental orchard of the Hesperides is bounded by olive trees around the perimeter. What is interesting is that Herakles and his bodyguard are depicted as *outside* that boundary. The apple tree, as one would expect, is centrally located in the orchard as a 'specimen tree' (complete with snake). It needs little observation to see that the apple leaves, trunk and growth habit

Fig. 7.3. The garden of the Hesperides, detail, on a krater by the Meidias Painter, (London, British Museum E224; Burn 1987: pl. 2c [M5]).

Fig. 7.4. Olives or myrtle pleached into an arch on a vase by the Meidias Painter, (Florence, Archaeological Museum 81947; Burn 1987: pl. 27a [M2]).

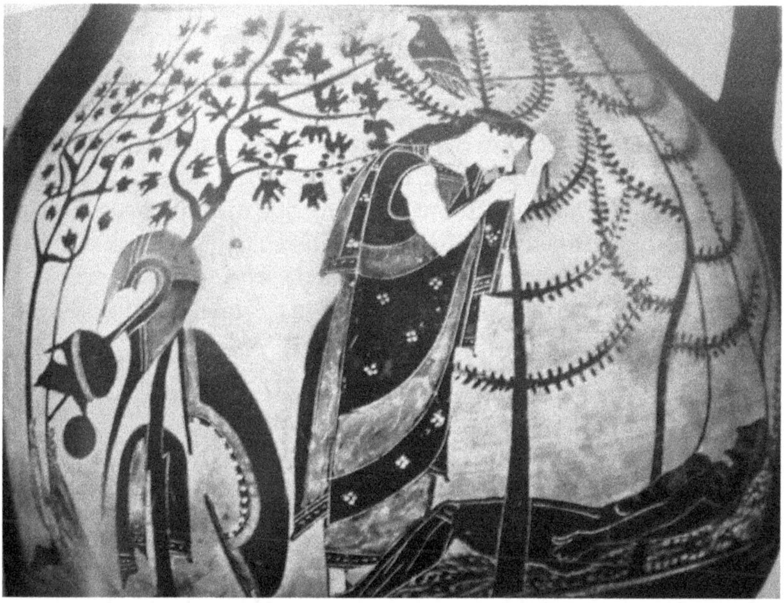

Fig. 7.5. Eos with the body of Memnon (?) in an orchard or grove, Attic amphora, sixth century BCE (Vatican 350).

are carefully distinguished from the olives within the conventions and techniques of vase painting. (Fig. 7.3) Indeed, a close look at the apple tree reveals that the top has been cut back and crown grafted. This is not a random rustic scene but a carefully constructed representation based on genuine formal and ornamental plantings of trees.

Other vases in the same style also show ornamental orchard scenes. The convention of using the tops of trees in the foreground to represent a perimeter marked by olive trees is found also in other vases of the same groups most notably the Judgment of Paris scene on Karlsruhe 259 (C1–Burn 1987: pl 40). Shrubby myrtle is sometimes shown (London E 698 [Burn P 1], Burn 1987: pl 20d). Another vase (Florence 81947 [Burn M2], Burn 1987: pl. 27a) shows two cordoned olives, or possibly myrtles, trained together (perhaps pleached) to form an arching bower (Fig. 7.4). Such bowers could be constructed of frameworks of withies. A fragment of Nikander preserved in

Athenaeus (15.683c) gives instructions, in verse, for training ivy over a frame in this way. Two vases may well show festoon-trained trees, (Cleveland 82.142 [M20]; Burn 1987: pl 12b and London GR 1895.10–29.2 [MM 96], Burn 1987: pl 12d).

Earlier vase painters also depict scenes in ornamental orchards and groves. In a vase by the Painter of the Vatican Mourner (Fig. 7.5) a woman is shown having laid out the body of a man, perhaps Eos and Memnon, in such a setting. Care has been taken to distinguish different types of trees: two plane trees are shown on the left and olives or possibly poplars are depicted on the right arranged in a formal planting pattern.

7.5. WOMEN IN ORCHARDS

There are also many scenes by a range of painters showing women picking fruit or playing in formally planted orchards (Fig. 7.6). These suggest that such ornamental orchards were considered to be acceptably safe places out-of-doors for women and girls, at least some of the time.

It is interesting that the places where ornamental vegetation grew were the locations considered most appropriate for women to pursue outdoor activities: the courtyard (an important working 'room' for women), the fountain house, and the orchard—where women are so often shown in activities somewhere between relaxing and working. Even at ornamentally planted tomb sites women might claim to have legitimate business as those who looked after the dead (Fig. 7.5). Why was this so? Perhaps the screening provided by well-tended ornamental vegetation was felt to 'protect' women in both a physical as well as a cosmic sense. Was a tamed orchard nearly as safe as a tamed house, or considered in some way to be an extension of it?

But these venues have overtones of danger as well: note the rape of the Leukippides and the comic eroticism of the passage from the *Acharnians.* Orchards could thus also be erotic locations (Osborne 1992b: 387), as in the black figure depiction by the Acheloos Painter (Fig. 7.7) of an embracing couple under a leafy arch of vines. Did the qualities which were thought to make orchards and similarly planted locations safe for women, such as seclusion and privacy, along with the very presence of women, make them both safe and dangerous simultaneously?

Fig. 7.6. Girls playing on a see-saw in an ornamental orchard, fragment of an Attic column krater (Boston Museum of Fine Arts 10.191).

Fig. 7.7. Couple embracing under a leafy arch of vines on an Attic pelike (London British Museum W40).

7.6. PLANTS IN FOUNTAIN HOUSES

Another venue for ornamental planting which appears in the vase painting evidence is fountains and fountain houses (Fig. 7.8). These are almost invariably shown with vegetation growing around and through them. Often the foliage seems to be vines, but also perhaps planes, poplars, willows, and other water-hungry trees also seem to have been planted around fountains and springs. Even if some of these could have grown 'naturally', without being planted on purpose (though I doubt that would be the case for vines), they would still have to be pruned and tended if the structure and its water pipes were not to be undermined by tree roots, and the entryways were not to become obstructed by branches and foliage. Unwanted trees, suckers,

Fig. 7.8. Women in a fountain house with vegetation on an Attic hydria (London British Museum B329).

and saplings would have to be removed. It seems certain from the way vegetation is depicted in so many of these scenes that it was considered appropriate and desirable for fountain houses.

7.7. PLANTS AND TREES IN HOUSES AND BUILDINGS

Within and around classical Greek houses, plants were clearly used ornamentally in several different ways. The most obvious use of plants within houses was the vine growing in the courtyard up the colonnade and perhaps across a trellis which features in so many Greek houses in one form or another. Many of these colonnades are south facing. Hence the vine would provide welcome screening and shade in the summer, but would let in light and warmth in the winter when the plant had lost its leaves. Examples appear in vase people dining under vines, in scenes which are probably depicting just such an arrangement of a trellis-grown vine in a courtyard (Figs. 7.9 and 7.10), rather than dining altogether out of doors. Free-standing trees also appear in courtyards (Fig. 7.11), as in the depiction of a perfumer's shop, perhaps selling his wares in the courtyard of his workshop in a set up reminiscent of the later houses with workshops at Halieis and Olynthos (Chapter 6.3.3; 6.3.4). And, there appears to be a free-standing tree at the door to the underworld, under which Kerberos sits to guard the entrance (Fig. 7.12)

A number of pit features identified in courtyards or near the doorways of houses and occasionally associated with other kinds of buildings (such as the *hestiatorion* attached to the submerged Temple of Apollo at Halieis, Jameson 1974: 115) may also have been used for planting trees and vines to shade courtyards and doorways from the hot summer sun. The features which have been identified by Ault (1994a, 1999) as '*koprones*' within the courtyards of Halieis houses form a good starting point (Figs. 7.13; 6.10; 7.14). The pits in Houses A, D and 7, ranged in size from less than 1 m to 2.5 m per side, and varied in area from 2.25 to 5 m^2, with a cubic capacity of 3–5 m^3.

Fig. 7.9. Herakles dining in a courtyard planted with a vine, on an Attic belly amphora (Munich 2301).

Fig. 7.10. Diners under a trellis-grown vine in a courtyard on an Attic cup (Oxford, Ashmolean Museum 1974.344).

Fig. 7.11. Sale of perfume or oil under a tree in a courtyard on an Attic pelike (Vatican 413).

Fig. 7.12. Kerberos under a tree at the entrance to the Underworld with Herakles and Athena on an Attic amphora (Paris Louvre F 204).

Their depth was generally 1 to 1.5 m in to the earth below the adjacent floor levels. Although they were lined with stone slabs they were not watertight and there were no traces of hydraulic plaster (Ault 1999: 550). At the time of excavation they were filled with earth which had a high proportion of artefactual (particularly ceramic) material. Some on the upper levels clearly comes from levels associated with collapse and abandonment, but sherds are distributed throughout the fill (Ault 1999: 550, 552). The example in House 7 had a drain running into it from the street (Ault 1999: 554).

Ault (1994a, 1999), following Owens (1983), has identified the use of these 'koprones' as receptacles for domestic waste 'contributing to

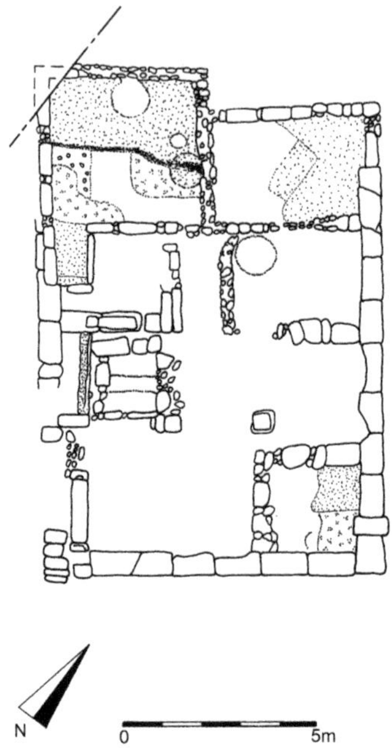

Fig. 7.13. Halieis, House A (after Ault 1999: 551, fig. 3).

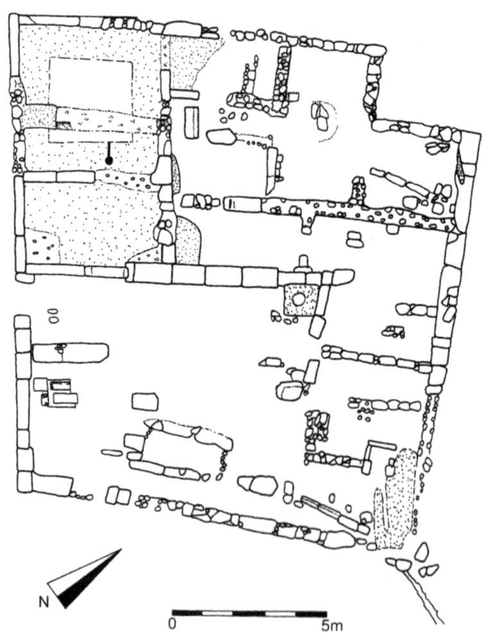

Fig. 7.14. Halieis, House 7 (after Ault 1994b).

a profitable compost heap or mulch pile' (Ault 1999: 557). He envisages the contents as including: 'animal and vegetal matter such as kitchen refuse and table scraps, and waste materials from domestic industrial activity, as well as human and animal excrement' (Ault 1999: 556). He also suggested that these 'compost pits' were emptied regularly and their contents removed to be spread on fields. Hence, he argues that the bulk of the large amounts of sherd material incorporated in the fill of these features dates to the last weeks or months of the houses' occupation (Ault 1999: 550, n. 5).

There are several serious problems with this identification. The first problem is the contents. Domestic waste such as by-products of food processing and 'table scraps' (if they even constituted a significant category of waste) were almost certainly fed to domestic animals (Foxhall 1998c). If these pits contained human and animal excrement, as he supposes, their location inside houses in domestic work areas seems highly unlikely. We know that Greeks of the classical period regularly segregated activities and materials which were considered 'polluting' in symbolic and other senses, among which excretion and excrement feature regularly (e.g. Hes. *WD* 727–32; 757–9). Ault's (1999: 557, n. 23) defence that modern sensibilities should be put aside and that the unpleasant smell would not necessarily been offended ancient Greeks in this respect seems weak, since living in close proximity to polluting substances would certainly have been offended by them. Apart from the smell, such repositories would have attracted annoying flies and other insects in large numbers, many of which bite, and this would have made working in such locations unbearable much of the time. Moreover, if they did contain excrement, why (as in House 7) have a drain flowing into the pit from the street, rather than the other way around?

The second problem is that there is simply too much sherd material for 'normal' household waste which is periodically removed. For two of these pits at Halieis, exisiting artefact assemblages have been analysed. The pit in House 7 contained 900 sherds (minimum of 230 vessels) and 300 roof tile fragments. That in House D contained over 1500 sherds (minimum of 144 vessels) and over 1000 roof tile fragments (some of which were clearly from the collapsed roof of the building) (Ault 1999: 552–4 and App.1, 567–8). If this is the domestic refuse which accumulated over the last 'weeks or months' of occupation, then this seems an extraordinarily high consumption and breakage rate for domestic pottery: how long would it take for a household, or even a business,[1] to break 100 or 200 pots? On the other hand, the amount of identifiable organic material was minimal, which seems peculiar, if these really served as 'compost heaps'.

The identification of similar features in the courtyards of the fourth century phases of houses in Athens (H. Thompson 1959: 10–1–2 and pl. 19, 21) as 'cesspits' runs up against similar objections. In the Athenian case the stone-lined pits are located in courtyards and immediately outside doors in the street, and again at least one (H. Thompson 1959: pl. 21; Owens 1983: 47) has a drainpipe leading *into* the pit (not out of it). As in the case of Halieis, the pits contained large quantities of sherd material. Both Owens (1983: 47) and H. Thompson (1959: 102) imply that these pits might have been used as latrines. This seems unlikely given the lack of privacy of their locations, either open to view in a courtyard, or even more exposed in a public street.

On the other hand, the locations of all of these pits are perfect for a tree or vine planted in a courtyard or in front of a door, and trained or pruned to provide useful shade. In the courtyard of House A at Halieis, a vine planted in the pit could be easily trained across the adjacent south facing portico, providing shade for the courtyard in summer, but allowing the sun to penetrate the work area (including a pressing room) in winter. This may also be the case in House D, though less of the courtyard was excavated. In House 7 the pit could have contained either a free-standing tree in the courtyard or a vine trained up a roughly west facing portico to the east of the pit (two stone bases appear to be in place). Similarly, the pits in the excavated Athenian houses are all conveniently located to provide a shade tree or vine at the door of a house or in a courtyard.

It is clear from the vase painting evidence discussed above that trees were a regular feature of courtyards and doorways. Surely dining in the courtyard would be preferably to a stuffy dining room in hot weather. Dem. 47.55 also depicts the women and children of a family having a meal in the courtyard.

[1] House 7 at Halieis may well be a 'taverna' in addition to being a domestic residence, see Foxhall 2007; Kelly-Blazeby 2006.

The presence of trees or vines planted in these pits account for several of their odd features. First of all, the arrangements for conveying water into these pits are easily explained by the presence of a tree and perhaps other plants. Secondly, it is likely that when the courtyard was swept, the planting pit for a tree or vine no doubt provided a handy hole for dumping sweepings. It is possible that manure was occasionally added to these planting pits. These practices, alongside the possibility that sherds were deliberately incorporated with the soil to assist drainage (important in winter even in dry climates, especially if the soil is clayey), may explain both the large amounts of pottery and the large minimum number of vessels which these pits contained. If a tree or large vine were planted in such a pit, then sherds could accumulate over time, because they would *not* be regularly removed. The presence of such pits suggests that we

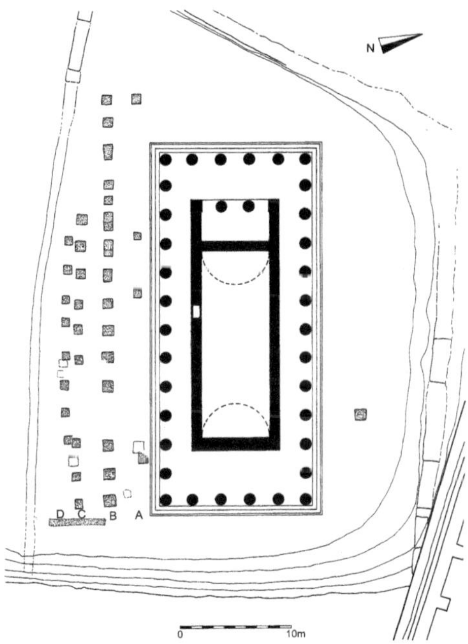

Fig. 7.15. Plan of the Hephaisteion in Athens showing planting pits for garden (D. Thompson 1937: 399, fig. 2; courtesy of the Trustees of the American School of Classical Studies at Athens).

should imagine classical Greek houses as incorporating trees and perhaps other plants into their architectural space and design.

The sanctuary garden of the Hephaisteion in Athens, probably dating to the early third century BCE (D. Thompson 1937: 410), included vines and trees, but plants in containers were also an important feature for at least part of its life. The planting holes at the Hephaisteion were cut into the bedrock and arranged in four rows, apparently constructed at different times, on the south side of the sanctuary between the temple building and the boundary wall (Fig. 7.15). This would have created a shady garden in summer on what would have been the hottest and sunniest part of the site, though if deciduous trees were used, the sun would still warm the area in winter. Water for the garden may have been provided by a catchment basin just outside the southwestern side of the temple precinct (D. Thompson 1937: 398).

In size, the planting holes are a little smaller than the Halieis '*koprones*': the largest (in row B) average 0.9 m per side, and they range in depth from 0.65–0.9 m. Several of those in row B, which was probably the original row, seem to have been enlarged at a later stage (D. Thompson 1937: 404). Significantly, like their counterparts at Halieis, they contained abundant sherd material when excavated (D. Thompson 1937: 410). Originally it appears that vines or trees were planted directly in these holes, but in the latest phases of the garden at least some of the holes were used as settings for flower pots. The features at Halieis, set in earth though lined with stone to restrict root spread, allowed space for trees or large vines to grow without permitting roots to become intrusive. It may be that the enlargement of some holes and the later use of flower pots indicates that the planting holes at the Hephaisteion were not particularly successful for larger trees or vines in the longer term. With holes cut into bedrock, drainage may have been a problem, especially in the winter. The site slopes upward to the south, and the holes may have filled with rainwater draining from the Kolonos Agoraios hill above. However, in summer, depending on the depth of soil above the holes, there may have been problems with plants drying out.

The use of flower pots in the later phases suggests the holes were used for shrubs or small vines, rather than for large vines or trees. The flower pots from the Hephaistion are around 20 cm high—about the same

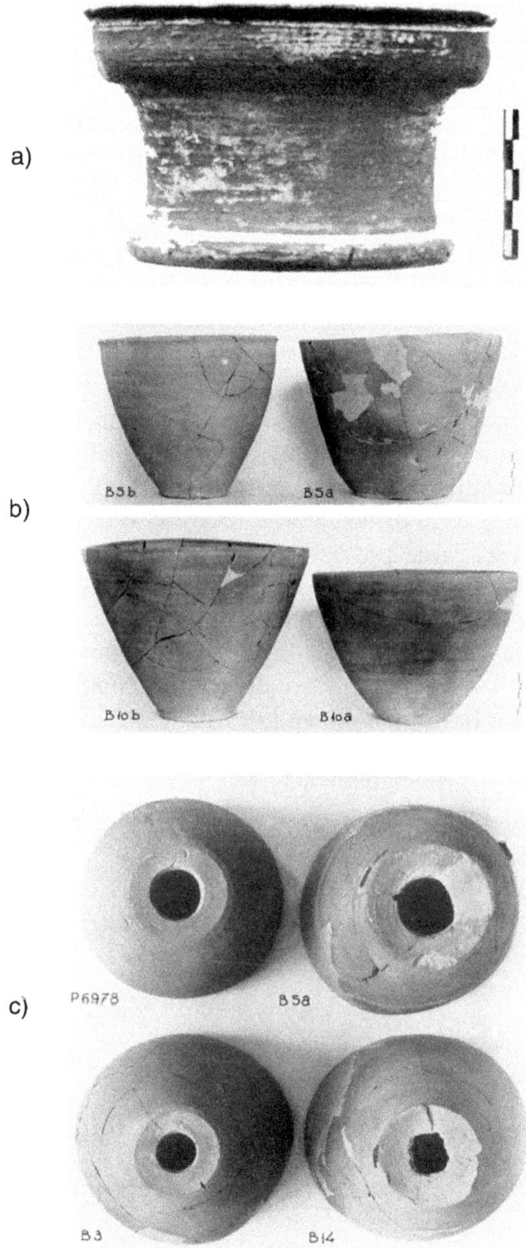

Fig. 7.16. Flower pots: a) Olynthos (Thompson 1937: 409, fig. 11); b) and c) Hephaisteion, Athens (Thompson 1937: 407, 405; figs. 7, 9 and 11; courtesy of the Trustees of the American School of Classical Studies at Athens).

size pot as are now recommended for many container-grown plants like herbs, or even smallish tomatoes. A handful of flowerpots were also identified in various houses in Olynthos (Robinson and Graham 1938) (Fig. 7.16), and it is possible that more may have been missed. Many flower pots in ordinary homes may have consisted of cracked or broken domestic pottery and hence their secondary use as plant containers may not be recognizable. The Olynthos pots are smaller than those from the Hephaistion—about half the size, and would have been suitable for herbs or small flowers. Certainly cracked and broken pots were recycled: this is clear from those recorded as sold in the Attic Stelae (e.g. the *φιδάκνε(ς) στόμα* of *IG* I^3 425.37—a bottomless vessel which could have been used as a flower pot, among other things).

There is some evidence from Theophrastos of the kinds of plants grown in containers. In large pots trees might be planted (*CP* 1.12.9)—if they were to be allowed to grow to any size they would have to be very large pots! The most obvious plants to grow in containers in a town garden may have been culinary herbs—oregano, basil, coriander, parsley, thyme, mint, and so forth, and there is some literary evidence for this (Nikander: Athenaios 15.684ab). But flowers, and indeed flowering culinary and medicinal herbs like rue, might also have been grown in containers.

Fig. 7.17. Attic pyxis with wedding scene showing cut foliage in pots (London British Museum E 774).

Cut flowers and foliage were a feature of wealthy households and were essential for entertaining and special occasions. The pyxis shown in Fig. 7.17 depicting a wedding scene shows not only garlanded women but also familiar Greek pottery shapes in use as flower vases. Rooms and even pots could be decorated with vegetation. Most often this is foliage—we are back to those ornamental orchards to provide the olive, ivy, myrtle, and the various flowers for decoration.

7.8. THE VALUE OF ORNAMENTAL ORCHARDS AND GARDENS

Ornamental orchards had both economic and social value. Obviously, only the rich could mobilize the skilled labour and capital to establish and maintain such plantings. Clearly they profited from them in several different ways. The economic and social value of the fruiting crops was probably enhanced by their attractive and elaborate planting—it seems likely that exceptionally 'good quality' was attributed to such produce. In addition, some of these gardens appear to have specialized in producing high-quality table fruit and flowers which would have fetch a high price if surpluses were available for sale. The sale value of the purely ornamental plants these orchards contained may often have been minimal, and any profits made must usually have been on quite a small scale, highly seasonal and, when they occurred, something of a windfall. Though crops from what may have been small ornamental orchards were sold by the *poletai* and recorded on the Attic Stelae, no flowers are mentioned. This does not mean they were not there, but could imply that they were not in season. Most of the edible and non-edible ornamental plants in ornamental orchards must have been grown for the consumption of wealthy households themselves, and for use in entertaining. Though surpluses may have been sold, the gains in money were probably relatively small. However, the gains in terms of prestige and the scope for building identities and relationships by lavishing exquisite garlands or choice table fruits on guests is likely to have been much more important. Such orchards then become part of the self-construction of the elite through the consumption of their special products.

8

Conclusions

The central theme of this work has been the ways in which the cultivation of the olive fits into both the cultural and the agricultural systems of the Mediterranean region of ancient Greece, as part of 'the economy'. Agriculture is one kind of cultural appropriation of the natural world (Bourdieu 1977; 1993: 214–25). In the case of complex societies like those of ancient Greece and Rome, such appropriation is via economic and political, as well as social, institutions (cf. Thomas 1983: 254–6 for the phenomenon in seventeenth- and eighteenth-century England). Hence, the study of farming and farming techniques, as I have shown, is as inseparable from the study of these wider realms as it is indispensable to the understanding and reconstruction of them. This leaves us with the difficult question of whether to study the chicken or the egg, or even whether we can really be sure which it is we are studying. The fact is, of course, it must be both. The relationship of chicken to egg is both processual and dynamic, like the political and social appropriation of nature that we call the agricultural economy. To study the agronomic technology of a single crop in isolation, without looking at the larger realms of which it is a part is like, to maintain the metaphor, trying to learn about egg production from a supermarket frozen chicken: isolated in its plastic wrapping, all the vital connections to support a proper explanation have been cut off. In that sense, this work forms an extended case study, focusing on a single crop and its exploitation as a means of understanding the larger contexts into which it is integrated.

For this reason it has proven necessary to consider a great many aspects of life in classical Greece beyond the cultivation of the olive. It is also why many traditional approaches to agronomic technology rapidly reach the limits of their interpretative capacity (White 1984 and many others). Remarkably often technological developments (or the lack of them) have been isolated from the overall cultural, social, economic, and political contexts with which they are entwined. Why, for example, should ancient Greeks have bothered with harrows when in most cases the operation itself was unnecessary because of the high labour input of other parts of the cultivation process? No more were dedicated, 'high-tech' pressing establishments appropriate in agricultural regimes where no one unit cultivated huge numbers of olive trees (Amouretti 1986: 107–8 and Chapter 6) because flexibility and opportunism were perceived to be more important to long-term economic and social success than maximizing production. The explanations depend as much on labour and organization, the transfer of wealth, soils, and climatic conditions as on a knowledge of tools or press types. The relatively small-scale farming enterprises of the classical Greek elite, at least in part constrained by the political boundaries of the polis and the limits on land ownership beyond one's own polis in most times and places in the classical Greek world, were as much a product of 'the economy' in its wider environmental setting as a cause of it. This work has also tried to explore 'the ancient economy' from another angle. To pick up chickens and eggs again, modern observers of 'the ancient economy' have often tried to study egg production by dissecting the omelette. And though not a bad way to learn about cookery, this does not tell you where the eggs came from.

The olive has a different 'life span' from that of an individual person or any particular *oikos*; a man plants them aiming to feed his grandchildren, though such long-term plans might often never come to fruition. The households we can see most clearly from such a great temporal distance are those of the wealthy whose motivations were plainly quite unlike those of the 'traditional peasant'. These households struggled to balance the conflicting demands and values of short-term opportunism in a volatile political and economic environment against the long-term necessity, to ensure that the *oikos* could satisfactorily reproduce itself by creating new *oikoi* in the next generation. The terminology of modern economics and the short-term reasoning of modernist 'economic rationalism' have not been equipped to cope with such strategies. Hence, it is difficult to use modern economic terms of analysis for interpreting the long-term view of an *oikos*, or a city founded in *oikoi*. For the Athenians especially (though perhaps not exclusively), the olive came to symbolize these ideals of the long-term: 'the leaf of the sweet, child-nurturing olive', *glaukas paidotrophou fullon elaias* (Soph. *OC* 694–706).

The household has continued to be a place where 'the economy' in some sense resides in our own world, despite the fact that it was encompassed by the hierarchies of the modern nation-state. In 1822, when the Industrial and Agricultural Revolutions were hatching out, and another comparable transformation in economic relationships was on the verge of happening, William Cobbett wrote his *Cottage Economy* partly in anticipation of the Act of Parliament that would make allotments available to the 'labouring classes', as he called them. Cobbett's ideal was lifted from Xenophon's:

> The word Economy, like a great many others, has, in its application, been very much abused. It is generally used as if it meant parsimony, stinginess, or niggardliness; and, at best, merely the refraining from expending money, hence misers and close-fisted men disguise their propensity and conduct under the name of economy; whereas the most liberal disposition, a disposition precisely the contrary of that of the miser, is perfectly consistent with economy.
>
> Economy means management and nothing more; and it is generally applied to the affairs of a house and family, which affairs are an object of the greatest importance, whether as relating to individuals or to a nation. A nation is made powerful and to be honoured in the world, not so much by the number of its people as by the ability and character of that people; and the ability and character of a people depend, in a great measure, upon the economy of the several families, which, all taken together, make up the nation. There never yet was, and never will be, a nation permanently great, consisting, for the greater part, of wretched and miserable families
>
> (Cobbett 1979: 1–2)

The Enlightenment celebration of the individual, for Cobbett at least, did not overshadow the fundamental organising principle of the household. He reiterates the classical ideal of self-sufficiency, *autarkia*, an ideal whose relevance lies in the fact that 'economy' is to some extent shaped by and embedded within 'household'; in Cobbett's case the very households which were to supply the commoditized labour for the mills.

However, in the Greek households we have examined in this work, self-sufficiency was less a reality than a mentality of self-containment—in terms of their economic activities these households, like the city-states of which they were a part, preferred to look inward rather than outward, but that does not mean that they were in reality economically or socially atomized. In perceiving the environment in which they operated as volatile, unstable and sometimes hostile, they preferred to keep their options open. In consequence, flexibility and opportunism characterized their choices and goals. At many levels ranging from the forging of political friendships and social alliances to the building of equipment for agricultural processing the capacity to turn something to several different uses, or to construct in such a way as to allow for change was desirable.

Nonetheless, it is plain that for the wealthy households best documented in the written and archaeological sources such self-containment entailed production at levels far beyond 'subsistence' on a regular basis. I have explored this as 'domestic' production (section 2.2.2), in effect one of several means by which these self-contained households competitively engaged with others via material goods. The large estates of Italy during the Republic and the earlier Principate provide an obvious comparative test case for exploring the utility of the concept of 'domestic production' and highlighting the rather different relationships between household and economy in the Roman world.

The prominence of large estates and their place in the structure of the Italian countryside of Roman times has been much debated from many different points of view (White 1967; 1970; Dyson 1978; 1992; 2003; 1981: 272; Frederiksen 1970–1; 1980; Duncan-Jones 1976; Garnsey 1998; Skydsgaard 1980; Spurr 1986; Rathbone 1981; Kolendo 1980; Giardina and Schiavone 1981; Wickham 1988; Kehoe 1997; Horden and Purcell 2000: 282–4). The relatively abundant source material from the agronomic treatises of Cato, Varro and Columella, the letters of Cicero and Pliny the Younger, inscriptions, and the evidence provided by archaeological survey and excavation, have raised questions such as the amount of the available agricultural land large estates swallowed up, the amount of total agricultural production for which they were responsible, and their relationships with small-scale cultivators, especially in terms of diachronic changes in those relationships. The answers to such questions must profoundly affect our reconstruction and understanding of the agrarian economies of Roman Italy.

It is plain despite the glum outlook of Plutarch (Plut. *T. and C. Gracchus* 8.2–7) on the predominance of large estates that small-scale cultivators continued to operate alongside large-scale property owners, and that even independent 'peasant' households still survived to eke out an existence right to the end of the Empire (Garnsey 1988; 1998; Dyson 1992). The evidence provided by archaeological survey (Dyson 1978; 2003; Barker and Lloyd 1991; Potter 1979; Barker and Hodges 1981 H. Patterson 2004; for methodological issues see Mattingly and Witcher 2004) suggests that there may have been quite large numbers of comparatively small farms. These must have been intimately connected at many levels to their larger neighbours (Garnsey 1998; de Neeve 1984; Foxhall 1990). It is probable that smaller-scale farmers grew different crops differently from their larger neighbours because they had different economic and social priorities and fewer resources. It is thus also likely that they used less elaborate and cheaper technologies than their larger neighbours, but under the guise of several possible relationships could have provided labour for these larger neighbours at busy times of year (Brunt: 1971: 194; Garnsey 1998; de Neeve 1984; J. Patterson 1987; Jongman 1988). On the other hand, it is also clear that by the later Republic much of central and southern Italy was in the hands of large landholders, especially land which has been appropriated by the state as *ager publicus* in the recent past, and land which was not easily exploited for intensive farming without the large-scale outlay of resources (Brunt 1988).

Large Roman estates appear to have been very large compared with their predecessors in Greece. The most comprehensive account of known estate sizes remains that of Duncan-Jones (1982: 323–6). Outside Egypt (e.g. Bowman 1985), there are few records of the sizes of 'real' estates. One of 200 *iugera* (51 ha) and several of 1000 *iugera* (253 ha) are known; also mentioned are a vineyard of 60 *iugera*

(15 ha) and another of probably 360 *iugera* (91 ha) (Duncan-Jones 1982: 324–5). Farm units of 100 or 200 *iugera* (25 or 51 ha) are given as examples of ideal properties by the agronomists (Cato *RR* 1.7; Varro *RR* 1.19.1; Columella 2.12.7). The *olivetum* of 240 *iugera* (61 ha) and the vineyard of 100 iugera (25 ha) could be actual sizes of properties owned by Cato himself, although Duncan-Jones (1982: 325–6) suggests that many land holdings were much larger, citing the area around Leontini in Sicily where 30,000 *iugera* (7590 ha) of wheat were in the hands of 32 proprietors. As in the Greek world, large-scale landholders frequently owned estates consisting of several independently operating farm units, sometimes in different regions. These units were themselves often very large and could be composed of non-contiguous parcels of land. Certainly the descriptions of ideal landholdings provided by the agronomic writers (Cato *RR* 1.7; Varro *RR* 1.6; Columella 1.2.3–5) stress the desirability of owning a range of different kinds of land which would be unlikely to have been co-terminous.

In contrast, it is noteworthy that the Romans at least believed that the traditional size of a peasant holding in Italy was 7 *iugera* (1.77 ha) (Evans 1980: 159, 161). If this figure bears any relationship to reality for any period, taken in conjunction with the documented sizes of many of the allotments given to veterans during the late Republic (Brunt 1988; Keppie 1983), it suggests that a number of farmers had holdings too small to support a household (cf. White 1970: 345–6). Whatever the reality it is clear that the difference in size between the largest and smallest properties was enormous. Wage labour, tenancy, patronage or some other tie of dependence to a wealthier neighbour may have offered the only means of survival for some small-scale farmers (Brunt 1988; J. Patterson 1987; Foxhall 1990).

The notion of 'domestic consumption' and the framework of the household are arguably relevant to understanding the agrarian economies of the Roman world. But, there are significant differences. The most obvious is one of scale. The household of a Pliny (Kehoe 1988; 1989; 1993) dwarfs that of a Phainippos (de ste Croix 1966). Nonetheless, the principle on which it functioned as a framework containing the economic 'enterprises' and activities of a kin-based group and their dependants, and the platform which underpinned economic activities more generally, were similar. Pliny's 'household' contained his various properties, houses and other economic activities (e.g. Plin. *Ep*. 7.11). However, these were themselves often self-contained and detachable units, as in the property set aside to support his 'charity' (*alimenta*) for the support of orphans (Plin., *Ep*. 7.18.2–3), or the *fundus* ('farm') set aside for the support of his old nurse (Plin., *Ep*. 6.3). Units could be changed or joined together, as in the case of the property adjoining one of Pliny's, which, if he were to buy it, he intended to merge with his existing holding (Plin., *Ep*. 3.19). Even in the case of property set aside to generate charitable funding, the property did not take on a new corporate status, but remained part of Pliny's household, with the income generated by the property devoted to the *alimenta*.

At a different level we have already shown that agricultural processing in Roman times was both far more specialized and greater in scale—permanent installations for the separate processing of grapes and olives virtually never, if ever, appear before the Principate, but were a relatively common feature of villas and large 'farms' throughout the world of the Roman Mediterranean. So, although Roman economies could be conceived as contained within households, in practical terms the scale of operations, including agricultural processing, as well as in the scope of the physical arena in which they were played out, had become so large that by the early Principate the household had become a tight fit for the economic activities stuffed into it. Within Greek poleis larger and smaller households of free citizens (and to some extent metics) appear to have operated as modular units, sitting side by side, but aspiring to look inward. Although they were politically contained within the state, economically and socially the state had little coercive control over them. In other words, their relationships to each other and to the state were mechanical, not organic.

In contrast, the political and social relationships which underpinned Roman societies were quite different in character from those of classical Greece. The difference is manifest in the contrast between Cicero's *De Officiis* 1.53–5 and one of its models, Aristotle's *Politics* (1252a24–b27; 1253b1–13). For Aristotle:

> It was out of the association formed by men with these two, women and slaves, that a household (*oikia*) was first formed ... This association of

persons, established according to nature for the satisfaction of daily needs, is the household...

The next stage is the village, the first association of a number of houses for the satisfaction of something more than daily needs. It comes into being... as offshoots of a household are set up by sons and grandsons.... This is why states were at first ruled by kings, as are foreign nations to this day: they were formed from constituents which were themselves under kingly rule. For every household is ruled by its senior member, as by a king, and the offshoots too, because of their blood relationship, are ruled in the same way.

The final association, formed of several villages, is the state (*polis*). For all practical purposes the process is now complete; self-sufficiency (*autarkia*) has been reached, and while the state came about as a means of securing life itself, it continues in being to secure the *good* life.

(Arist. *Pol.* 1252b.)

Cicero, however, tracks the expansion of households and their links to others in much more detail, the sum of which, in his view, eventually forms the foundation for the state:

For since the reproductive instinct is by Nature's gift the common possession of all living creatures, the first bond of union is that between husband and wife; the next that between parents and children; then we find one household (*domus*) with everything in common. And this is the foundation of civil government, the nursery, as it were, of the state. Then follow the bonds between brothers and sisters, and next those of first and then second cousins; and when they can no longer occupy one household, they go out into other households as into colonies. Then follow between these, in turn, marriages and connections by marriage, and from these again a new stock of relations; and from this propagation and after-growth states (*rei publici*) have their beginnings. (Cicero *de Officiis* 1.54.)

Far from building conceptual walls around an (ideally) self-contained unit, wealthy households in the Roman world looked outward to a much greater extent and deliberately sought connections with other households, however fragile these might sometimes turn out to be. Beyond the family, many of these links were socially and politically asymmetrical. Patronage held a position of prominence in Roman social and political relationships that it never attained (though it certainly existed) in archaic and classical Greek societies.

The household as the location for economic activities was not redundant under Roman rule, but the concept of the household itself and its place in larger political and social structures appears transformed. The practice of seeking external links in the context of enhanced social hierarchies encouraged large households to encompass (or at least try to do so) those of lower status to which they were linked by relations of power. The upshot was a complex chain of networked hierarchical relationships which could sometimes be largely political and economic in character. Plainly the wealthiest Roman households could grow so large, and spread themselves over such a wide geographical area, that genuine personal relationships were no longer feasible. Nonetheless, the organizational model of the 'household' was retained until the end of Rome, despite its transformation into a completely different kind of political entity, organically connected to other households, with the imperial household holding the prime place at the top of the hierarchy, conceptually, at least, encompassing the whole of the empire.

For wealthy Roman households producing at levels far beyond 'subsistence', 'domestic' consumption remains at one level a relevant concept. The high moral value continued that to be placed on the ideal of 'self-sufficiency' (Duncan-Jones 1982: 37) attests to this. If encompassment was at the heart of hierarchy, to improve the position of one's own household, it was essential in reaching outward to contain others rather than to be contained, to swallow instead of being swallowed. In practice the economic resources of a household, and any surpluses generated by economic activities, could be used to transform these material resources into social and political relationships which themselves became sources of further economic gain. The result may be something that looks like 'growth' in a modern economic sense in terms of the generation of surpluses and 'profits', but which has emerged from institutional structures and mentalities which are not those of the modern world. Ultimately, the retention of the household as the structural model for the location of many economic activities must have limited the scope for the development of the kinds of free-standing economic institutions we find in the modern industrial and post-industrial world. Nonetheless, the Roman emphasis on making connections, and the concomitantly expanded scale of economic activities in the context of empire, permitted space for economic activities, relationships and institutions beyond the household, absent in the in the world of classical Greece.

Lack of technological innovation and 'progress' in classical antiquity has been a major theme from Finley (1999: 146–8; 1965; Saller 2002) onward. At the most general level this is attributed to cultural and political values inherent in ancient societies. Finley's view remains typical today:

> 'There was never a time, so far as I know, when the large landholders of antiquity did not prosper as a class.... their psychology was that of the rentier, and hence neither their material circumstances nor their attitudes were favourable to innovation. They were not so stupid or so hidebound... that they could not (sometimes) tell a better landed investment from a poorer one. But essentially their energies went into spending their wealth, not making it, and they spent it on politics and the good life.'
>
> (Finley 1983b: 188–9.)

For classical Greece, didactic works about the practicalities of farming have not survived. Even if they did it is unlikely that innovation and experimentation would be issues specifically confronted. They are not major concerns, in an introspective way, of the sources we do have. However, that does not mean that innovation and experimentation were lacking. The existing botanical books of Theophrastos (*CP* 3.2.4) give the impression that curiosity was implemented without being considered remarkable enough to warrant comment, though this may reflect his philosophical perspective. But, experimentation must have been the prerogative of the wealthy: poorer farmers (or craftsmen) could not afford to take the risks of failure.

I hope that I have shown in this work that whether technological innovations were adopted or not was a complex combination of matter of practicality and mentality. The practical considerations were as often shaped by the natural as by the political and social environments. The large-scale adoption of innovation not only entails risk, and the ability of even the wealthiest land owners of the Classical period to confront high risks was limited, but may also have limited flexibility. The latter could well have been a very important consideration for householders operating in environments perceived to be unstable and volatile, where maintaining flexibility and opportunism was a key aim. The really significant difference is the pathway of innovation is trodden in 'small steps for (woman or) man' rather than in 'great leaps for mankind'.

Many new inventions are improvements precisely because they allow the same processes to be carried out on a larger scale. But, as I have shown in the case of the rotary olive crusher, there is no advantage to increased through-put if the rest of the system does not produce more to put through. Necessity really is the mother of invention insofar as its adoption must go hand in hand with systemic change. As long as households and poleis remained relatively self-contained (thought not necessarily self-sufficient) entities, and while flexibility at all levels was a paramount consideration, the incentive to expand the scale and intensity case of any particular element within a fairly wide-ranging mixed farming regime was lacking. The expense in labour and resources of improved equipment, any time lag on returns, and any concomitant decrease in flexibility, could have been enough to discourage the adoption of new machines by most farmers until its utility was beyond doubt. That archaeologically discernable presses become more common and the rotary olive crusher becomes well established in the period when Macedonian imperial ambitions begin to change the political landscape of the polis is therefore probably no accident. This is confirmed by the adoption of these technological innovations on a much wider scale in Roman times when, as discussed above, a completely different conceptualization of 'household' as the seat of economic activities, in combination with the broad geographical contexts of conquest and empire, combined to encourage agricultural units which were both more specialized and producing on a larger scale than in the world of Classical poleis. These developments profoundly changed the agrarian landscape of Greece and other parts of the Mediterranean region, and their effects are still visible today in the archaeological record. This process also suggests that both the agrarian landscapes of Greece and the socio-cultural values associated with its products had changed in meaning. The olive and the vine, and the techniques of their cultivation and processing, therefore become emblematic of these large scale transformations, though the crops and technologies *per se* did not initiate or cause them.

On the other hand, it is clear that certain commodities, in combination with specifically elaborated techniques of cultivation and production, come to have high cultural and symbolic values. It could be argued that this is a phenomenon which, at a most basic level, forms the foundation for all behaviours which we broadly classify as 'economic' in human societies. Conferring value upon an object reifies it, creating a materiality it did not necessarily possess intrinsically. That is, an object in this sense could be a thing, but it could also be a service, idea, technique, relationship, image, or anything else, which through the cultural and social processes by which they come to be valued, are transformed in effect into concrete objects with a material existence which can be measured against something else. The specifics of the cultural and social processes of desire, through which such values are conferred, are almost infinitely variable. So too the specific impact of such valuation must vary with social, political and cultural context. This is not simply the classic notion of supply and demand in another guise, for we are looking here at a phenomenon more fundamental and lower-level than market mechanisms, which functions in economies or sectors of economies with money. Desire, valuation and the consequent reification of the desired object are social acts which define the relationships of people and the things in their world.

For person or a group to confer value on an object implies the desire to acquire, possess, and in a general sense, consume it. Consumption is thus entwined with social relationships and hierarchies, and simultaneously produces culturally-specific behaviours for engaging with the material world and appropriating it into human society. Consumption becomes fulfilment of desire; the validation of valuation. But it can also be much more than this: the materiality conferred upon valued objects, their commodification as it were, enables them to be accumulated and exchanged to fulfil further desires generated by groups and individuals within specific social hierarchies. Desire itself is, of course, culturally specific, but generally it has little to do directly with 'need' in any absolute sense. At the very least, what any particular society considers to be 'necessary' is shaped by and buried within broader cultural and social parameters.

So what is the difference between then and now? What distinguishes our economies from theirs? The degree to which specific desires and attribution of values link into wider social ambitions and hierarchies must be a significant part of the explanation. But it is possible that the process of commodification itself assists in the creation of hierarchies, or at least fuels or reinforces extant hierarchies, encouraging their expansion and elaboration via shared desires and values. In a sense, how effectively commodification enhances social and political manoeuvrability depends on the scale at which others understand, engage with, and share to some extent the values attributed to specific objects. (Arguably, the wider the pool of shared values and the closer the degree of sharing, the more effective the use of consumption to validate, enhance or even change identities.) The key issues for spotting the differences between us and them then becomes why the social locations where value is conferred on objects have expanded and how the range of institutions through which desire is fulfilled via consumption have come into being.

Elaborations of desire in regard to food offer a particularly good example of these social practices; as summarized by Garnsey (1999: 108), 'food behaviour reflects the social hierarchy and social relationships'. At one level practicalities such as seasonality, limited storage life, and most especially, perceived notions of 'quality', become key elements in culturally-based valuations. Food and its consumption are therefore not simply a matter of sustenance or subsistence, but serve as a statement of personhood in a particular context, a material manifestation of social identities. As a basic principle, this is as true for the modern world as for classical antiquity, but the institutions and relationships through which it is expressed are different.

The consumption of such commodities, their literal embodiment, generally within socially significant contexts, forms part of the iterative process of engagement with the material world by which individuals and households construct their identities. The place of such commodities in the social and cultural order thereby becomes entwined with ritual (*sensu lato*) and public behaviours and a sense of appropriateness. In classical antiquity olive oil is plainly one of the most important of these. In classical Greece this is shown by its elaborated consumption as *opson* or 'relish' with food, for bodily cleansing in the gymnasium and adornment in the form of perfume, and for lighting. Wine, of course, is another. In Roman times many of these same values continue to be attributed to olive oil, but the geographical spread of these values and the number of people who engaged with them in one way or another had dramatically expanded. Plainly the Roman world produced far larger quantities of olive oil than classical Greece, but this does not of course mean that it was equally distributed in geographical or social terms among the pool of those desiring it. The Monte Testaccio (Blazquez 1992) in Rome provides a material testimony of the desire of the city's inhabitants to consume olive oil, the values conferred upon it, and the ability of Rome to fulfil that desire to consume (and even to distribute it to, and thus fulfil the desires of, the relatively poor), manifesting in the process the supremacy of Rome as the centre of a vast empire. This is a long way from the polis communities of archaic and classical Greece, where the primary arenas for commodification, consumption, and, for the most part, ambitions, were largely (though certainly not exclusively) contained within individual city-states. In other words, the exclusivity of classical Greek poleis, maintained by competition between them, mitigated against the development of larger territorial and institutional arenas in which the ambitions of wealthy individuals and families might expand.

At the heart of the extended case study pursued here, the exploitation of the olive, stand the farmers and their households. Which specific techniques they chose from those available in the larger cultural repertoire to grow the trees and to make the many of types of oil and table olives differently valued and consumed in classical antiquity depended heavily on why and how value was conferred upon the olive and its products in specific times and places, and how these products were commoditised in particular social and political contexts. Simultaneously, the processes and the limits of commodification must also have modified the techniques and practices of arboriculture chosen by farmers and explored here. The activities of these Classical farmers and their households in part shaped the landscapes of the Mediterranean which we see today. And this brings us back to chickens and eggs (Plut. *Mor.* 635E).

Bibliography

Acheson, P. E. (1997). 'Does the "economic explanation" work? Settlement, agriculture and erosion in the territiory of Halieis in the Late Classical–Early Hellenistic Period', *Journal of Mediterranean Archaeology*, 10: 165–90.

Adam-Veleni, P. and Mangafa, M. (1996). '*Αρχαίο ελαιοτρειβείο στα Βρασνά νομού Θεσσαλονίκης*', in S. Pappadopoulos (ed.) *Elaia kai Ladi*. Athens: ETBA (Industrial Development Bank of Greece), 92–104.

Alcock, S. (1993). *Graecia Capta: The landscapes of Roman Greece*. Cambridge: Cambridge University Press.

Allbaugh, L. G. (1953). *Crete: A case study of an underdeveloped area*. Princeton: Princeton University Press.

American School of Classical Studies (1953). *The Athenian Agora*. Princeton: The American School of Classical Studies.

Amouretti, M.-C. (1986). *Le pain et l'huile dans la Grèce anciene*. Paris: Les Belles Lettres.

—— and Brun, J.-P (eds.) (1993). *La production du vin et de l'huile en Méditerranée*. Paris: French School at Athens.

——, Comet, G. Ney, C., and Paillet, J. C. (1984). 'A propos d'un pressoir a huile: de l'archeologie industrielle a la histoire', *Melanges de l'Ecole Française de Rome, Antiquites*, 96(1): 379–421.

Amyx, D. A. (1958). 'The Attic stelae, part III', *Hesperia*, 27: 163–310.

Anderson–Stojanovič, V. (1996). 'The University of Chicago Excavations in the Rachi Settlement at Isthmia, 1989', *Hesperia*, 65: 57–98.

—— (forthcoming). *The Rachi Settlement*. Princeton: American School of Classical Studies.

André, J. (1964). 'La résine et la poix dans l'antiquité: technique et terminologies', *Antiquité Classique*, 33: 86–97.

Andreau, J. (1995). 'Vingt ans d'après l'economie antique de Moses I Finley', *Annales: Histoire, Sciences Sociales*, 50: 947–60.

Andreyev, V. N. (1967). 'Atticheskoe obshchestvennoe zemelevladenie V–III vv. do n. e', *Vestnik Drevnii Istorii*, 2 (102): 32–48.

Ault, B. A. (1994). *Classical Houses and Households: An Architectural and Artifactual Case Study from Halieis, Greece*. Ph.D. thesis, Department of Classics: Indiana University.

Ault, B. A. (1994). 'Koprones and oil presses: domestic installations related to agricultural productivity and processing at classical Halieis', in P. N. Doukellis and L. G. Mendoni (eds.) *Structures Rurales et Société Antiques.* Paris: Les Belles Lettres, 197–206.

—— (1999). 'Koprones and oil presses at Halieis: interactions of town and country and the integration of domestic and regional economies', *Hesperia*, 68: 549–73.

Ayala, G. and French, C. (2005). 'Erosion modelling of past land-use practices in the Fiume di Sotto di Torina river valley, north-central Sicily', *Geoarchaeology*, 20: 149–67.

Bakhuisen, S. C. (1991). 'Torcula Graecanica: a note on the archaeology of olive and grape pressing', in M. Gnade (ed.) *Stips votiva. Papers presented to C.M. Stibbe.* Amsterdam: Allan Pierson Museum Series: 1–5.

Barker, G. and Hodges, R. (eds.) (1981) *Archaeology and Italian Society.* Oxford: British Archaeological Reports.

—— and Lloyd, J. (eds.) (1991) *Roman Landscapes: archaeological survey in the Mediterranean region.* London: British School at Rome.

Benaki Museum (1978). *Paradosiakes kalliergeis.* Athens: Benaki Museum.

Bent, J. T. (1885). *The Cyclades or Life among the Insular Greeks.* London: Longmans, Green and Co.

Bintliff, J. (2005). 'Human impact, land use history and the surface archaeological record: a case study from Greece', *Geoarchaeology*, 20: 135–47.

—— Farinetti, E., Howard, P., Sarri, K., and Sbonias, K. (2002). 'Classical farms, hidden prehistoric landscapes and Greek rural survey: a response and an update', *Journal of Mediterranean Archaeology*, 15: 259–65.

Bintliff, J. and Snodgrass, A. (1985). 'The Cambridge/Bradford Boeotian Expedition: the first four years', *Journal of Field Archaeology*, 12: 123–61.

—— —— (1988). 'Mediterranean survey and the city', *Antiquity*, 62: 57–71.

Blackman, D. (1998). 'Archaeology in Greece 1997–98', *Archaeological Reports* 44: 1–128.

Blazquez, J. M. (1992). 'The latest work on the export of Baetican olive oil to Rome and the army', *Greece and Rome*, 39: 173–88.

Blok, A. (1974). *The Mafia of a Sicilian Village, 1860–1960.* Oxford: Blackwell.

Blok, J. and Lardinois, A. (eds.) (2006). *Solon of Athens: New historical and philological approaches.* Leiden: Brill.

Blümner, H. (1912). *Technologie und Terminologie der Gewerbe und Künste bei Griechen und Römern.* Leipzig and Berlin: Teubner.

Boardman, J. (1958–9). 'Excavations at Pindakas in Chios', *Annual of the British School at Athens*, 53–4: 295–309.

—— (1974). *Athenian Black Figure Vases.* London: Thames and Hudson.

—— (1975). *Athenian Red Figure Vases. The Archaic period.* London: Thames and Hudson.
—— (1989). *Athenian Red Figure Vases. The Classical period.* London: Thames and Hudson.
Bommeljé, S., Doorn, P., Deylius, M., Vroom, J., Bommeljé, Y., Fagel, R., and van Wijngaarden, H. (1987). *Aetolia and the Aetolians: Towards the Interdisciplinary Study of a Greek Region.* Utrecht: Parnassus Press.
Bonias, Z. and Perreault, J. (1993). '*Άργιλος, ἀνακσκαφή* 1992–3', *Archaiologiko Ergo ste Makedonia kai Thrake*, 7: 465–73.
—— —— (1997). '*Άργιλος, ἀνακσκαφή* 1997', *Archaiologiko Ergo ste Makedonia kai Thrake*, 11: 539–48.
Boserup, E. (1965). *The Conditions of Agricultural Growth.* London: Allen and Unwin.
—— (1981). *Population and Technology.* Oxford: Blackwell.
Boulotis, C. (1996). '*Το λάδι στους μυκηναϊκούς χρόνους*', in S. Papadopoulos (ed.) *Elaia kai ladhi.* Athens: ETBA (Industrial Development Bank of Greece), 19–58.
Bourdieu, P. (1977). *Outline of a Theory of Practice.* Cambridge: Cambridge University Press.
—— (1993). *The Logic of Practice.* Cambridge: Cambridge University Press.
Bowkett, L. (1995). *The Hellenistic Dye-Works.* Well Built Mycenae. The Helleno-British excavations within the citadel of Mycenae, 1959–69, vol. 36. Oxford: Oxbow.
Bowman, A. (1985) 'Landholding in the Hermopolite nome in the fourth century AD', *Journal of Roman Studies*, 75: 137–63.
Boyd, T. D. and Rudolph, W. W. (1978). 'Excavations at Porto Cheli and vicinity, preliminary report IV: the lower town of Halieis, 1970–1977', *Hesperia*, 47: 333–55.
Bradford, J. (1956). 'Fieldwork on aerial discoveries in Attica and Rhodes, II', *Antiquaries Journal*, 36: 172–80.
—— (1957). *Ancient Landscapes: Studies in Field Archaeology.* London: Bell.
Broneer, O. (1955). 'Excavations at Isthmia', *Hesperia*, 24: 110–41.
—— (1958). 'Excavations at Isthmia', *Hesperia*, 27: 1–37.
Brun, J.-P. (1986). *L'oléiculture antique en Provence. Les huileries du département du* Var. Paris: Éditions du Centre National de la Recherche Scientifique.
—— (1999). '*Lautissimum fuit antiquitus in Delos insula*: La maison IB du quartier du Stade et la fabrication des parfums à Délos à l'époque hellénistique', *Bulletin de Correspondance Hellenique*, 87–155.
—— (2000). 'The production of perfumes in antiquity: the cases of Delos and Paestum', *American Journal of Archaeology*, 104: 277–308.

—— (2003). *Le vin et l'huile dans la Méditerranée antique. Viticulture, oléiculture et procédés de transformation.* Paris: Editions Errance.

—— (2004a). *Archéologie du vin et de l'huile de la préhistoire à l'époque hellénistique.* Paris: Éditions Errance.

—— (2004b). *Archéologie du vin et de l'huile dans l'Empire romain.* Paris: Editions Errance.

—— (2005). *Archéologie du vin et de l'huile en Gaule romaine.* Paris: Editions Errance.

—— and Brunet, M. (1997). 'Une huilerie du premier siècle avant J.-C. dans le Quartier du théâtre à Délos', *Bulletin de Correspondance Hellenique*, 121: 573–615.

Bruneau, P. (1979). 'Les jardins urbains de Délos', *Bulletin de Correspondance Hellenique*, 103: 89–99.

Bruneau, P. and Fraisse, P. (1981). 'Un pressor à vin à Délos', *Bulletin de Correspondance Hellenique*, 105: 127–53.

—— —— (1984). 'Pressoirs déliens', *Bulletin de Correspondance Hellenique*, 108: 713–30.

Brunet, M. (1990a). 'Contribution à l'histoire rurale de Délos aux époques classique et hellenistique', *Bulletin de Correspondance Hellenique*, 114: 669–82.

—— (1990b). 'Terrasses de cultures antiques: l'exemple de Délos, Cyclades', in M. Provansal (ed.) *L'agriculture en terrasses sur les versants méditerranéens: histoire, consequences sur l'evolution du milieu.* Aix-en-Provence: Centre National de la Recherche Scientifique, 5–11.

—— (1992). 'Campagnes de la Grèce antique: les dangers du prisme athénien', *Topoi*, 2: 33–51.

—— (1999). 'Le paysage agraire de Délos dans l'antiquite', *Journal des Savants*, 1–50.

—— and Poupet, P. (1997). 'Territoire délien', *Bulletin de Correspondance Hellenique*, 121: 776–81.

——, Rougemont, G., and Rousset, D. (1998). 'Les contrats agraires dans la Grèce antique. Bilan historiographique illustré par quatres exemples', *Histoire et societé rurales*, 9: 211–45.

Brunt, P. (1971) *Italian Manpower 225* BCE*–AD 14.* Oxford: Clarendon Press.

—— (1972). 'Review of K.D. White, Roman Farming,' *Journal of Roman Studies*, 62,: 153–8.

—— (1988) 'The army and the land in the Roman revolution', in *The Fall of the Roman Republic and Related Essays.* Oxford: Clarendon Press, 240–80. Rev. ed. of article published under same title in *Journal of Roman Studies* (1962), 52: 69–86.

Burford, A. (1977). 'The family farm in Greece', *Classical Journal* 73: 162–75.

—— (1993). *Land and Labor in the Greek World.* Baltimore and London: Johns Hopkins University Press.

Burn, L. (1987). *The Meidias Painter.* Oxford: Oxford University Press.

Cahill, N. (2001). *Household and City Organization at Olynthus.* New Haven and London: Yale University Press.

Campbell, J. and Sherrard, P. (1968). *Modern Greece.* London: Benn.

Camps-Fabrer, H. (1953). *L'olivier et l'huile dans l'Afrique romaine.* Algiers: Direction de l'Interieur et des Beaux-Arts.

Carroll-Spillecke, M. (1989). *ΚΕΠΟΣ: Der antike griechischen Garten.* Munich: Deutscher Kunstverlag.

Carter, J. C. (1990). 'Metapontum–land, wealth and population', in J.-P. Descoeudres (ed.) *Greek Colonists and Native Population: Proceedings of the first Australian congress of classical archaeology held in honour of Emeritus Professor A. D. Trendall, Sydney, 9–14 July 1985.* Canberra and New York: Humanities Research Centre & New York: Oxford University Press, 405–41.

—— (2006). *Discovering the Greek Countryside at Metaponto.* Ann Arbor: University of Michigan Press.

Cartledge, P. (1998). 'The economy (economies) of ancient Greece', *Dialogos*, 5: 4–24.

——, Cohen, E. E., and Foxhall, L. (eds.) (2002). *Money, Labour and Land: Approaches to the economies of ancient Greece.* London: Routledge.

Cartledge, P., Millett, P., and von Reden, S. (eds.) (1998). *KOSMOS: Essays in Order, Conflict and Community in Classical Athens.* Cambridge: Cambridge University Press.

Casanova, A. (1993). 'Types de pressoirs et types de productions à partir de l'exemple de la Corse à la fin du XVIIIe siècle', in M.-C. Amouretti and J.-P. Brun (eds.) *La production du vin et de l'huile en Méditerranée.* Paris: French School at Athens, 359–78.

Catling, H. (1983–4). 'Archaeology in Greece', *Archaeological Reports*, 30: 3–70.

—— (1984–5). 'Archaeology in Greece', *Archaeological Reports*, 31: 3–69.

—— (1985–6). 'Archaeology in Greece', *Archaeological Reports*, 32: 1–101.

—— (1986–7). 'Archaeology in Greece', *Archaeological Reports*, 33: 3–61.

Cavanagh, W., Crouwel, J., Catling, R .W. V., and Shipley, G. (1996). *Continuity and Change in a Greek Rural Landscape. The Laconia Survey, vol. II, Archaeological Data,* London: British School at Athens.

Cayeux, L. (1911). *Exploration archéologique de Délos. Description physique de l'ile de Délos.* Paris: Fontemoing.

Charre, R. and Couilloud-Le Dinahet, M. T. (1999). 'Sites de fermes à Rhénée', in *Territoires des cités grecques. Actes de la table ronde internationale organisée par 'École française d'Athènes, 31 octobre–3 novembre 1991*. Paris: De Boccard, 135–57.

Chayanov, A. V. (1986). 'Peasant farm organization (1925)', in D. Thorner, B. Kerblay and R. E. F. Smith (eds.) *A.V. Chayanov on the Theory of Peasant Economy*. Manchester: Manchester University Press, 29–269.

Cherry, J. F. (1983). 'Frogs around a pond', in D. R. Keller and D. W. Rupp (eds.) *Archaeological Survey in the Mediterranean Area*. Oxford: BAR, 37–416.

Cherry, J. F. and Davis, J. L. (1985). 'The Nemea Valley Survey', *American Journal of Archaeology*, 89: 327.

—— —— (1986). 'The Nemea Valley Survey', *American Journal of Archaeology*, 90(90): 204–5.

—— —— (1987). 'The Nemea Valley Survey', *American Journal of Archaeology*, 91: 320.

—— —— (1991). 'The Ptolemaic base at Koressos on Keos', *Annual of the British School at Athens*, 86: 9–28.

—— —— and Mantzourani, E. (1984). 'The Kea Survey', *National Geographic Society. Research Reports*, 21: 109–116.

—— —— —— (1991). *Landscape Archaeology as Long-Term History: Northern Keos in the Cycladic Islands*. Los Angeles: University of California Press.

Cobbett, W. (1979). *Cottage Economy*. Oxford: Oxford University Press.

Cohen, E. E. (1992). *Athenian Economy and Society: A banking perspective*. Princeton: Princeton University Press.

Cox, C. A. (1998). *Household Interests: Property, marriage strategies, and family dynamics in ancient Athens*. Princeton: Princeton University Press.

Dahl, R. (1989). *Democracy and its Critics*. New Haven: Yale University Press.

Davidson, J. (1998). *Courtesans and Fishcakes: The consuming passions of classical Athens*. London: Fontana.

Davies, J. K. (1971). *Athenian Propertied Families*. Oxford: Clarendon Press.

—— (1981). *Wealth and the Power of Wealth in Classical Athens*. New York: Arno.

de Neeve, P. (1984). *Colonus: Private farm tenancy in Roman Italy during the Republic and Early Principate*. Amsterdam: North-Holland.

de Ste Croix, G. E. M. (1966). 'The estate of Phaenippus', in E. Badian (ed.) *Ancient Society and Institutions. Studies presented to V. Ehrenberg*. Oxford: Blackwell, 109–14.

—— (1981). *The Class Struggle in the Ancient Greek World*. London: Duckworth.

Deutsches Archaologisches Institut (1939–). *Kerameikos: Ergebnisse der Ausgrabungen*. Berlin: DAI.

Docter, R. (1991). 'Athena vs Dionysos, reconsidering the contents of SOS amphorae', *Bulletin Antieke Beschaving*, 66: 45–50.

Doukellis, P. N. (1994). 'Le territoire de la colonie romaine de Corinthe', in P. N. Doukellis and L. G. Mendoni (eds.) *Structures Rurales et Sociétés Antiques*. Paris: Les Belles Lettres, 391–6.

—— and Mendoni, L. G. (eds.) (1994). *Structures Rurales et Société Antiques*. Paris: Les Belles Lettres.

Drachmann, A. G. (1932). *Ancient Oil Mills and Presses*. Copenhagen: Levin and Munksgaard.

DuBoulay, J. (1974). *Portrait of a Greek Mountain Village*. Oxford: Oxford University Press.

Dufková, M. and Pečírka, J. (1970). 'Excavations of farms and farmhouses in the *chora* of Chersonesos in the Crimea', *Eirene*, 8: 123–74.

Duncan-Jones, R. (1976) 'Some configurations of landholding in the Roman Empire', in M. I. Finley (ed.) *Studies in Roman Property*. Cambridge: Cambridge University Press, 7–34.

—— (1982) *The Economy of the Roman Empire*. 2nd ed. Cambridge: Cambridge University Press.

Dyson, S. (1978) 'Settlement patterns in the Ager Cosanus', *Journal of Field Archaeology*, 5: 251–68.

—— (1992) *Community and Society in Roman Italy*. Baltimore and London: Johns Hopkins University Press.

—— (2003) *The Roman Countryside*. London: Duckworth.

Elbaum, R., Melemed-Bessudo, C., Boaretto, E., Galili, E., Lev-Yadun, S., Levy, A., and Weiner, S. (2006). 'Ancient olive DNA in pits: preservation, amplification and sequence analysis', *Journal of Archaeological Science*, 33: 77–88.

Erdkamp, P. (2005). *The Grain Market in the Roman Empire: A social, political and economic study*. Cambridge: Cambridge University Press.

Evans, J. K. (1980) '*Plebs rustica*: the peasantry of classical Italy', *American Journal of Ancient History*, 5: 19–47; 134–93.

FAO/WHO (Food and Agriculture Organization/World Health Organization) (1973). *Energy and Protein Requirements*. Geneva: WHO.

Finley, M. I. (1951). *Studies in Land and Credit in Ancient Athens, 500–200 B.C.* New Brunswick, NJ and Oxford: Rutgers University Press.

—— (1983a). 'Land, debt and the man of property in classical Athens', in R. Saller and B. Shaw (eds.) *Economy and Society in Ancient Greece*. Harmondsworth: Penguin, 62–76.

—— (1983b). 'Technical innovation and economic progress in the ancient world', in R. Saller and B. Shaw (eds.) *Economy and Society in Ancient Greece*. Harmondsworth: Penguin, 176–95.

Finley, M. I. (1999). *The Ancient Economy* (3rd edn). Introduction by Ian Morris. Berkeley and London: University of California Press.

Forbes, H. A. (1976). '"We have a little of everything" the ecological basis of some agricultural practices in Methana, Trizinia', in M. Dimen and E. Friedl (eds.) *Regional Variation in Modern Greece and Cyprus: Toward a perspective on the ethnography of Greece,* New York: New York Academy of Sciences, 236–50.

—— (1982). *Strategies and Soils: Technology, production and environment in the peninsula of Methana, Greece.* Ph.D. thesis, Department of Anthropology. Philadelphia: University of Pennsylvania.

—— (1989). 'Of grandfathers and grand theories: the hierarchised ordering of responses to hazard in a Greek rural community', in P. Halstead and J. O'Shea (eds.) *Bad Year Economics. Cultural responses to risk and uncertainty.* Cambridge: Cambridge University Press, 87–97.

—— (1993). 'Ethnoarchaeology and the place of the olive in the economy of the Southern Argolid, Greece'. *La Production du Vin et de l'huile en Mediterranée.* M.-C. Amouretti and J.-P. Brun. Paris: Boccard, 213–26.

—— (1994). 'Pastoralism and settlement structures in ancient Greece', in P. N. Doukellis and L. G. Mendoni (eds.) *Structures Rurales et Société Antiques.* Paris: Les Belles Lettres, 187–96.

—— (1996). 'The uses of the uncultivated landscape in modern Greece: a pointer to the value of the wilderness in antiquity?', in G. Shipley and J. Salmon (eds.) *Human Landscapes in Classical Antiquity: Environment and culture.* London: Routledge, 68–97.

—— (forthcoming) *Meaning and Identity in a Greek Landscape: An archaeological ethnography.* Cambridge: Cambridge University Press.

—— and Foxhall, L. (1978). '"The queen of all trees": preliminary notes on the archaeology of the olive', *Expedition,* 21: 37–47.

—— and Foxhall, L. (1995). 'Ethnoarchaeology and storage in the ancient Mediterranean: beyond risk and survival', in J. Wilkins, D. Harvey and, M. Dobson (eds.) *Food in Antiquity.* Exeter: Exeter University Press, 69–86.

—— and Koster, H. A. (1976) 'Fire, axe and plow: human influence on local plant communities in the Southern Argolid', in M. Dimen and E. Friedl (eds.) *Regional Variation in Modern Greece and Cyprus: Toward a perspective on the ethnography of Greece.* New York: New York Academy of Sciences, 109–126.

Fouchard, A. (1993). 'Le statut des agriculteurs dans la cité grecque idéale au IVe av. J.-C.' *Revue des Études Grecques,* 106: 61–81.

Foxhall, L. (1989). 'Household, gender and property in classical Athens', *Classical Quarterly,* 39: 22–44.

—— (1990). 'The dependent tenant: land leasing and labour in Italy and Greece', *Journal of Roman Studies*, 80: 97–114.
—— (1992). 'The control of the Attic landscape, in B. Wells (ed.) *Agriculture in Ancient Greece*. Stockholm: Swedish Institute at Athens, 155–9.
—— (1993a). 'Farming and fighting in ancient Greece', in J. Rich and G. Shipley (eds.) *War and Society in the Greek World*. London: Routledge, 134–45.
—— (1993b). 'Oil extraction and processing equipment in classical Greece', in M.-C. Amouretti and J.-P. Brun (eds.) *La production du vin et de l'huile en Méditerranée*. Paris: Boccard, 183–200.
—— (1995). 'Bronze to iron: agricultural systems in Late Bronze Age and Early Iron Age Greece and the implications for social and political structures', *Annual of the British School at Athens*, 90: 239–50.
—— (1996). 'Feeling the earth move: soil retention on steep slopes in classical antiquity', in G. Shipley and J. Salmon (eds.) *Human Landscapes in Classical Antiquity: Environment and Culture*. London: Routledge, 44–67.
—— (1997). 'A view from the top: evaluating the Solonian property classes', in L. Mitchell and P. Rhodes (eds.) *The Development of the Polis in Archaic Greece*. London: Routledge, 113–36.
—— (1998a). 'Cargoes of the heart's desire: the character of trade in the archaic Mediterranean world', in N. R. E. Fisher and H. van Wees (eds.) *Archaic Greece: New Approaches and New Evidence*. Swansea and London: Duckworth and The Classical Press of Wales, 295–310.
—— (1998b). 'The politics of affection: emotional attachments in ancient Athens' in P. Cartledge, P. Millett and, S. von Reden (eds.) *KOSMOS: Essays in Order, Conflict and Community in Classical Athens*. Cambridge: Cambridge University Press, 52–67.
—— (1998c). 'Snapping up the unconsidered trifles: the use of agricultural residues in ancient Greek and Roman farming', *Environmental Archaeology*, 1: 35–40.
—— (2001). 'Response to David Pettegrew, "Chasing the classical farmstead"', *Journal of Mediterranean Archaeology*, 14: 216–22.
—— (2002). 'Access to resources in classical Greece: the egalitarianism of the polis in practice', in P. Cartledge, E. E. Cohen and, L. Foxhall (eds.) *Money, Labour and Land: Approaches to the economies of ancient Greece*. London: Routledge, 209–20.
—— (2003). 'Cultures, landscapes and identities in the Mediterranean world', in I. Malkin (ed.) *Mediterranean Paradigms, Mediterranean Historical Review* 18: 75–92.

Foxhall, L. (2004a). 'Field sports: engaging Greek archaeology and history', in E. Sauer (ed.) *Archaeology and Ancient History: Breaking down the boundaries.* London: Routledge, 76–84.

—— (2004b). 'Small, rural farmstead sites in ancient Greece: a material cultural analysis', in F. Kolb (ed.) *Chora und Polis: Methoden und Ergebnisse der historische Landeskunde.* Munich: Oldenbourg, 249–70.

—— (2005). 'Village to city: Staples and luxuries? Exchange networks and urbanization', in B. Cunliffe and R. Osborne (eds.) *Mediterranean Urbanisation.* London: British Academy, 233–48.

—— (2007). 'House clearance: Unpacking the 'Kitchen' in Classical Greece', in R. Westgate, J. Whitley and N. Fisher (eds.) *Building Communities: Houses in the Aegean.* London: British School at Athens, 233–42.

—— and Forbes, H. A. (1982). 'Sitometreia: the role of grain as a staple food in classical antiquity', *Chiron*, 12: 41–90.

Frankel, R. (1999). *Wine and Oil Production in Israel in Antiquity in Israel and Other Mediterranean Countries.* Sheffield: Sheffield Academic Press.

—— Avitsur, S., and Ayalon, E. (1994). *History and Technology of Olive Oil in the Holy Land.* Arlington VA., and Tel Aviv: Oléarius Editions and Eretz Israel Museum.

Frederick, C. D. and Krahtopoulou, N. (2000). 'Deconstructing agricultural terraces: examining the influences of construction methods on stratigraphy, dating and archaeological visibility', in P. Halstead and C. D. Frederick (eds.) *Landscape and Land Use in Post-glacial Greece.* Sheffield: Sheffield Academic Press, 79–94.

—— —— (2003). 'The history of terracing and its impact' UCL, Kythera Island Project. http://www.ucl.ac.uk/kip/ga_ terrace.php. (Accessed 31 October 2006).

Frederiksen, M. (1970/71). 'The contribution of archaeology to the agrarian problem in the Gracchan period', *Dialoghi di archeologia*, 4–5: 330–57.

—— (1980) 'Plinio il Vecchio e l'agricoltura in età imperiale romana: gli aspetti technici ed economici', in E. Gabba (ed.) *Tecnologia, economia e società nel mondo romano: atti del convegno di Como, 27–29 settembre 1979.* Como: Banca popolare commercio e industria, 81–98.

Gagarin, M. (2006). 'Legal procedure in Solon's laws', in J. Blok and A. Lardinois (eds.). *Solon of Athens: New historical and philological approaches.* Leiden: Brill, 261–75.

Galili, E., Stanley, D. J., Sharvit, J., and Weinstein-Evron, M. (1997). 'Evidence for earliest olive oil production in submerged settlements off the Carmel coast, Israel', *Journal of Archaeological Science*, 24: 1141–50.

——, Weinstein-Evron, M., Hershkovitz, I., Gopher, A., Kislev, M., Lernau, O., Kolska-Horwitz, L., and Lernau, H. (1993). 'Atlit-Yam: a prehistoric site on the sea floor of the Israeli coast', *Journal of Field Archaeology*, 20: 133–57.

Gallant, T. (1982). 'Agricultural systems, land tenure and the reforms of Solon', *Annual of the British School at Athens*, 77: 111–24.

—— (1991). *Risk and Survival in Ancient Greece.* Stanford: Stanford University Press.

Garnsey, P. (1985). 'Grain for Athens', in P. Cartledge and F. D. Harvey (eds.) *Crux. Essays presented to G.E.M. de Ste Croix on his 75th Birthday.* Exeter: Exeter University Press, 62–75.

—— (1988). *Famine and Food Supply in the Graeco-Roman World. Responses to Risk and Crisis.* Cambridge: Cambridge University Press.

—— (1998). *Cities, Peasants and Food in Classical Antiquity: Essays in social and economic history.* Cambridge: Cambridge University Press.

—— (1999). *Food and Society in Classical Antiquity.* Cambridge: Cambridge Unversity Press.

Gates, M. H. (1996). 'Archaeology in Turkey', *American Journal of Archaeology*, 100: 277–335.

Ghiannakaris, A. (1985). '*Πότε καὶ πώς ποτίζονται οἱ ἐλαιώνες*', *Agrotika*, 230: 10–14.

Giardina, A. and Schiavone, A. (eds.) (1981). *Società romana e produzione schiavistica.* Bari: Editori Laterza.

Gill, C., Postlethwaite, N., and Seaford., R. (eds.) (1998). *Reciprocity in Ancient Greece.* Oxford: Oxford University Press.

Granovetter, M. (2005). 'Comment on Liverani and Bedford', in J. G. Manning and I. Morris (eds.) *The Ancient Economy: Evidence and Models.* Stanford: Stanford University Press, 84–8.

Grove, A. T. and Rackham, O. (2001). *The Nature of Mediterranean Europe: An ecological history.* New Haven and London: Yale University Press.

Hadjisavvas, S. (1992). *Olive Oil Processing in Cyprus from the Bronze Age to the Byzantine Period.* Göteborg: P. Åströms Förlag.

Hall, J. (2007) *A History of the Archaic Greek World.* Oxford: Blackwell.

Halstead, P. (1987). 'Traditional and ancient rural economy in Mediterranean Europe: plus ca change?' *Journal of Hellenic Studies*, 107: 77–87.

—— and Jones, G. (1989). 'Agrarian ecology in the Greek islands: time stress, scale and risk', *Journal of Hellenic Studies*, 109: 41–55.

—— and O'Shea, J. (eds.) (1989). *Bad Year Economics. Cultural responses to risk and uncertainty.* Cambridge: Cambridge University Press.

—— —— (1989). 'Introduction: cultural responses to risk and uncertainty', in P. Halstead and J. O'Shea (eds.) *Bad Year Economics: Cultural responses to risk and uncertainty.* Cambridge: Cambridge University Press, 1–7.

Hamilakis, Y. (1996). 'Wine, oil and the dialectics of power in Bronze Age Crete: a review of the evidence', *Oxford Journal of Archaeology*, 15: 1–32.

—— (1999). 'Food technologies/technologies of the body: the social context of wine and oil production and consumption in Bronze Age Crete', *World Archaeology*, 31: 38–54.

Hansen, J. (1985). 'Palaeoethnobotany in Cyprus: recent research', *American Journal of Archaeology*, 89: 332–3.

—— (1988). 'Agriculture in the prehistoric Aegean: data versus speculation', *American Journal of Archaeology*, 92: 39–52.

Hansen, M. H. (1986). *Demography and Democracy: the Number of Athenian Citizens in the Fourth Century B.C.* Herning: Systime.

Hanson, V. D. (1992). 'Practical aspects of grape growing and the ideology of Greek viticulture, in B. Wells (ed.) *Agriculture in Ancient Greece.* Stockholm: Swedish Institute at Athens, 161–6.

—— (1998). *Warfare and Agriculture in Classical Greece* (rev. edn). Berkeley and London: University of California Press. Originally published (1983), Pisa: Giardini.

—— (1999). *The Other Greeks: The family farm and the agrarian roots of western civilization* (2nd edn). Berkeley and London: University of California Press Originally published (1995) New York: The Free Press.

Harris, E. (2002). 'Workshop, marketplace and household: the nature of technical specialization in classical Athens and its influence on economy and society', in P. Cartledge, E. E. Cohen, and L. Foxhall (eds.) *Money, Labour and Land: Approaches to the economies of ancient Greece.* London Routledge, 67–99.

Hartmann, H. T. and Opitz, K. W. (1977). *Olive Production in California.* Berkeley: Division of Agricultural Sciences, University of California.

Harvey, F. D. (1985). 'Dona ferentes: some aspects of bribery in Greek politics, in P. Cartledge and F. D. Harvey (eds.) *Crux. Studies in Honour of G.E.M. de Ste Croix on his 75th Birthday.* Exeter: Exeter University Press, 76–117.

Hitchner, R. B. (1993). 'Olive production and the Roman economy: the case of intensive growth', in M.-C. Amouretti and J.-P. Brun (eds.) *La production du vin et de l'huile en Méditerranée.* Paris: Boccard, 499–508.

Hjohlman, J., A. Penttinen, and Wells, B. (2005). *Pyrgouthi. a rural site in the Berbati valley from the early iron age to late antiquity.* Stockholm: Paul Åströms Förlag.

Hodder, I. and Hutson, S. (2003) *Reading the Past: Current approaches to interpretation in archaeology* (3rd edn). Cambridge: Cambridge University Press.

Hodkinson, S. (1986). 'Land tenure and inheritance in classical Sparta', *Classical Quarterly*, 36: 378–406.

—— (1988). 'Animal husbandry in the Greek polis', in C. R. Whittaker (ed.) *Pastoral Economies in Classical Antiquity. Proceedings of the Cambridge Philological Society,* supp. vol. 14. Cambridge: Cambridge Philological Society, 35–74.

—— (1989). 'Inheritance, marriage and demography: perspectives upon the success and decline of classical Sparta', in A. Powell (ed.) *Classical Sparta: Techniques behind her success.* Norman, OK: Oklahoma University Press, 79–121.

—— (1992). 'Sharecropping and Sparta's economic exploitation of the helots', in J. M. Sanders (ed.) *ΦΙΛΟΛΑΚΩΝ: Studies in Honour of Hector Catling.* London: British School at Athens, 123–34.

—— (2000). *Property and Wealth in Classical Sparta.* London and Swansea: Duckworth and The Classical Press of Wales.

Hoepfner, W. and Schwander, E.-L. (1994). *Haus und Stadt in klassischen Griechenland.* Munich: Deutscher Kunstverlag.

Horden, P. and Purcell, N. (2000). *The Corrupting Sea: A study of Mediterranean history.* Oxford: Blackwell.

Humphreys, S. C. (1980). 'Family tombs and tomb cult in ancient Athens: tradition or traditionalism', *Journal of Hellenic Studies,* 100: 96–126.

Hunter, V. (1989). 'Women's authority in classical Athens', *Échos du Monde Classique/Classical Views,* 8(1): 39–48.

—— (1994). *Policing Athens. Social control in the Attic lawsuits, 420–320 BCE.* Princeton: Princeton University Press.

Isager, S. and Skydsgaard, J.-E. (1992). *Ancient Greek Agriculture. An introduction.* London: Routledge.

Jameson, M. H. (1969). 'Excavations at Porto Cheli and vicinity, preliminary report I: Halieis 1962–8', *Hesperia,* 37: 311–42.

—— (1974). 'The excavation of a drowned Greek temple', *Scientific American,* 231: 111–19.

—— (1977). 'Agriculture and slavery in classical Athens', *Classical Journal,* 73(2): 122–145.

—— (2001). 'Oil presses of the Late Classical/Hellenistic period', in J.-P. Brun and P. Jockey (eds.) *Technai, Techniques et sociétés méditerranéennes. Homage à Marie-Claire Amouretti.* Paris: MMSH/Maisonneuve et Larose, 281–99.

——, C. N. Runnels, *et al.* (1994). *A Greek Countryside. The Southern Argolid from prehistory to the present day.* Stanford: Stanford University Press.

Jardé, A. (1925). *Les céréales dans l'antiquité grecque.* Paris: Boccard.

Johnson, T. and C. Dandeker (1974). 'Two Attic country houses', *Athens Annals of Archaeology* 7: 303–12.

Johnson, T. and C. Dandeker (1975). 'Town and country houses of Attica in classical times', in Mussche, H., Spitaels, P, Goemaere-De Poerck, F. (eds.). *Thorikos and the Laurion in Archaic and Classical Times. Miscellanea Graeca 1*. Ghent: Belgian Archaeological Mission in Greece, 63–140.

Johnston, A. W. and R. E. Jones (1978). 'The SOS amphora', *Annual of the British School at Athens*, 73: 103–42.

Jones, G. (1982). 'Cereal and pulse remains from Protogeometric and Geometric Iolkos, Thessaly', *Anthropologika*, 3: 75–8.

Jones, J. E., Graham, A. J., and Sackett, L. H. (1973). 'An Attic country house below the cave of Pan at Vari', *Annual of the British School at Athens*, 68: 355–452.

—— Sackett, L. H., and Graham, A. J. (1962). 'The Dema house in Attica', *Annual of the British School at Athens*, 57: 75–114.

Jones, N. F. (2000). 'Epigraphic evidence for farmstead residence', *Zeitschrift für Papyrologie und Epigraphik*, 133: 75–90.

—— (2004). *Rural Athens under the Democracy*. Philadelphia: University of Pennsylvania Press.

Jongman, W. (1988). *Economy and Society of Pompeii*. Amsterdam: Gieben.

Kardara, C. (1961). 'Dyeing and weaving at Isthmia', *American Journal of Archaeology*, 65: 261–66.

Kehoe, D. (1988) 'Allocation of risk and investment on the estates of Pliny the Younger', *Chiron*, 18: 15–42.

—— (1989) 'Approaches to economic problems in the 'letters' of Pliny the Younger: the question of risk in agriculture', *Aufstieg und Niedergang der Römischen Welt*, II.33.1: 555–90.

—— (1993) 'Economic rationalism in Roman agriculture', *Journal of Roman Archaeology*, 6: 476–84.

—— (1997) *Investment, Profit and Tenancy: The jurists and the Roman agrarian economy*. Ann Arbor: University of Michigan Press.

Keller, D. (1989). 'Classical Greek agricultural sites: the Karystian evidence', *American Journal of Archaeology*, 93: 275.

Keller, D. R. and Rupp, D. W. (eds.) (1983). *Archaeological Survey in the Mediterranean Area*. BAR International Series 195. Oxford: BAR.

Kelly-Blazeby, C. (2006). *Kapeleion: 'Casual' and commercial wine consumption in Classical Greece*. Ph.D. thesis, School of Archaeology and Ancient History, University of Leicester.

Kent, J. H. (1947). 'The temple estates of Delos, Rheneia and Mykonos', *Hesperia*, 17: 243–338.

Keppie, L. (1983) *Colonisation and Veteran Settlement in Italy, 47–14 BCE*. London: British School at Rome.

Kislev, M., Hartmann, A., and Bar-Josef, O. (2006). 'Early domesticated fig in the Jordan Valley', *Science*, 312(5778): 1372–4.

Kistler, E. (2004) 'Kampf der Mentalitäten: Ian Morris' 'elitist' versus middling-ideology'?', in Rollinger, R. and Ulf, C. (eds.) *Griechische Archaik: Interne Entwicklungen, externe Impulse.* Berlin: Akademie-Verlag, 145–75.

Kloner, A. and Sagiv, N. (1993). 'The olive presses of Hellenistic Maresha, Israel', in M.-C. Amouretti and J.-P. Brun (eds.) *La production du vin et de l'huile en Méditerranée.* Paris: Les Belles Lettres, 119–35.

Kolendo, J. (1980) *L'agricoltura nell'Italia romana.* Biblioteca di storia antica 10. Rome: Editori riuniti.

Koparal, E. and İplikçi, E. (2004). 'Archaic Olive Oil Extraction Plant in Klazomenai', in A. Moustaka, E. Skarlatidou, M.-C. Tzannes, and Y. E. Ersoy (eds.) *Klazomenai, Teos and Abdera: Metropoleis and Colony. Proceedings of the international symposium held at the Archaeological Museum of Abdera, 20–21 October 2001.* Thessaloniki: University of Thessaloniki, 221–234.

Koster, H. A. (1977). *The Ecology of Pastoralism in Relation the Changing Patterns of Land Use in the Northeast Peloponnese.* Ph.D. thesis, Department of Anthropology, University of Pennsylvania.

Krasilnikoff, J. (2002). 'Water and farming in classical Greece: evidence, method and perspective', in K. Ascani, V. Gabrielsen, K. Kvist, and A. H. Rasmussen (eds.) *Ancient History Matters. Studies presented to Jens-Erik Skydsgaard on his seventieth birthday.* Rome: Bretschneider, 47–62.

Kurke, L. (1991). *The Traffic in Praise: Pindar and the poetics of social economy.* Ithaca and London: Cornell University Press.

—— (1999). *Coins, Bodies, Games and Gold: The politics of meaning in archaic Greece.* Princeton: Princeton University Press.

—— (2002). 'Money and mythic history: the contestation of transactional orders in the fifth century BCE', in W. Scheidel and S. Von Reden (eds.) *The Ancient Economy.* Edinburgh: Edinburgh University Press, 87–113.

Lambert, S. D. (1997). *Rationes Centesimarum. Sales of public land in Lykourgan Athens.* Amsterdam: Gieben.

Landers, J. (1987). 'Mortality and metropolis: the case of London, 1675–1825', *Population Studies*, 41: 59–76.

—— and Mouza, A. (1988). 'Burial, seasonality and causes of death in London 1670–1819', *Population Studies*, 42: 59–83.

Langdon, M. (1990–91). 'On the farm in classical Attica', *Classical Journal*, 86: 209–13.

—— and Watrous, L. V. (1977). 'The farm of Timesios: rock-cut inscriptions in South Attica', *Hesperia* 46: 162–7.

Lewis, D. M. (1973). 'The Athenian Rationes Centesimarum', in M. I. Finley (ed.) *Problemes de la terre en Grece ancienne*. Paris: Mouton, 187–212.

Lloyd, J. A., Owens, E. J., and Roy, J. (1983). 'The Megalopolis Survey in Arcadia', in D. R. Keller and D. W. Rupp (eds.) *Archaeological Survey in the Mediterranean Area*. Oxford: BAR, 267–9.

Lohmann, H. (1985). 'Landleben in klassischen Attika', *Jahrbuch Ruhr-Universität Bochum*: 71–96.

—— (1992). 'Agriculture and country life in classical Attica', in B. Wells (ed.) *Agriculture in Ancient Greece*. Stockholm: Swedish Institute at Athens, 29–57.

—— (1993). *Atene: Forschungen zu Siedlungs- und Wirtschaftstruktur des klassischen Attika*. Cologne, Weimar, Vienna: Böhlau.

—— (1994). 'Ein "Alter Schafstall" in neuem licht: die Ruinen von Palaia Kopraisia bei Legrena (Attika)', in P. N. Doukellis and L. G. Mendoni (eds.) *Structures Rurales et Société Antiques*. Paris: Les Belles Lettres, 81–132.

Loomis, W. (1998). *Wages, Welfare Costs and Inflation in Classical Athens*. Ann Arbor: University of Michigan Press.

MacDonald, W. A. and Rapp, G. R. (1972). *The Minnesota Messenia Expedition: Reconstructing a Bronze Age regional environment*. Minneapolis: University of Minnesota Press.

Manning, J. G. and Morris, I. (eds.) (2005). *The Ancient Economy: Evidence and models*. Stanford: Stanford University Press.

Matson, F. R. (1972). 'Ceramic studies', in W. A. McDonald and G. R. Rapp (eds.) *The Minnesota Messenia Expedition. Reconstructing a Bronze Age regional environment*. Minneapolis: University of Minnesota Press, 200–24.

Mattingly, D. J. (1988a). 'Megalithic madness and measurement, or how many olives could an olive press press?' *Oxford Journal of Archaeology*, 7(2): 177–95.

—— (1988b). 'Oil for export? A comparison of Libyan, Spanish and Tunisian olive oil production in the Roman empire', *Journal of Roman Archaeology*, 1: 33–56.

—— (1988c). 'The olive boom. Oil surpluses, wealth and power in Roman Tripolitania', *Libyan Studies*, 19: 21–41.

—— (1993). 'Maximum figures and maximising strategies of oil production? Further thoughts on the processing capacity of Roman olive presses', in M.-C. Amouretti and J.-P. Brun (eds.) *La production du vin et de l'huile en Méditerranée*. Paris: Boccard, 483–98.

—— (1996a) 'Olive presses in Roman Africa: technical evolution or stagnation?', in M. Khanoussi, P. Ruggeri, and C. Vismara (eds.) *L'Africa*

Romana: Atti dell'convegno di studio XI, Cartagine, 15–18 Dicembre 1994. Ozieri: Editrice il Torchietto, 577–95.

—— (1996b). 'First fruit? The olive in the Roman world' in G. Shipley and J. Salmon (eds.) *Human Landscapes in Classical Antiquity: Environment and culture.* London: Routledge, 213–53.

—— and Hitchner, R. B. (1993). 'Technical specifications for some North African olive presses of Roman date', in M.-C. Amouretti and J.-P. Brun (eds.) *La production du vin et de l'huile en Méditerranée.* Paris: Boccard, 439–62.

—— and Salmon, J. (eds.) (2001). *Economies Beyond Agriculture in the Classical World.* London: Routledge.

—— and Witcher, R. (2004). 'Mapping the Roman world: the contribution of field survey data', in S. Alcock and J. Cherry (eds.) *Side-by-Side Survey. Comparative regional studies in the Mediterranean world.* Oxford: Oxbow, 173–86.

Mee, C. B. and Forbes, H. A. (eds.) (1996). *A Rough and Rocky Place: Settlement and land use in the peninsula of Methana, Greece.* Liverpool: University of Liverpool Press.

—— Gill, D., Forbes, H. A., and Foxhall, L. (1991). 'Rural settlement change in the Methana Peninsula, Greece', in G. Barker and J. Lloyd (eds.) *Roman Landscapes: Archaeological Survey in the Mediterranean Region.* London: British School at Rome, 223–32.

Melena, J. L. (1983). 'Olive oil and other sorts of oil in the Mycenaean tablets', *Minos*, 19: 89–123.

Miekle, S. (2002). 'Modernism, economics and the ancient economy', in W. Scheidel and S. Von Reden (eds.) *The Ancient Economy.* Edinburgh: Edinburgh University Press, 233–50.

Millett, P. (1983). 'Maritime loans and the structure of credit in fourth-century Athens', in P. Garnsey, K. Hopkins and C. R. Whittaker (eds.). *Trade in the Ancient Economy.* London: Chatto and Windus, 36–52.

—— (1984). 'Hesiod and his world', *Proceedings of the Cambridge Philological Society* 30: 84–115.

—— (1989). 'Patronage and its avoidance in ancient Athens', in A. Wallace-Hadrill (ed.) *Patronage in Ancient Society.* London: Routledge, 15–47.

—— (1991). *Lending and Borrowing in Ancient Athens.* Cambridge: Cambridge University Press.

Morel, J.-P. (2001). 'Aux origines de moulin rotatif? Une meule circulaire de la fin du 6eme siècle avant notre ère à Carthage', in J.-P. Brun and P. Jockey (eds.) *Technai, Techniques et sociétésen Méditerranéennes. Homage à Marie-Claire Amouretti.* Paris: MMSH/Maisonneuve et Larose, 240–51.

Morris, I. (1994a). 'The Athenian economy twenty years after The Ancient Economy', *Classical Philology*, 89: 351–66.

—— (ed.) (1994b). *Classical Greece: Ancient Histories and Modern Archaeologies.* Cambridge: Cambridge University Press.

—— (1996). 'The strong principle of equality and the archaic origins of democracy', in J. Ober and C. Hedrick (eds) *Dêmokratia: A conversation on democracy, ancient and modern.* Princeton: Princeton University Press, 19–48.

—— (2000). *Archaeology as Cultural History.* Oxford: Blackwell.

—— (2005). 'Archaeology, standards of living, and Greek economic history', in J. G. Manning and I. Morris (eds.) *The Ancient Economy: Evidence and models.* Stanford: Stanford University Press, 91–126.

Morris, S. and Papadopoulos, J. (2005). 'Greek towers and slaves: an archaeology of exploitation', *American Journal of Archaeology*, 109: 155–225.

Munn, M. and Munn, M. L. (1989). 'The Stanford Skourta Plain Project: the 1987 and 1988 seasons of survey on the Attic-Boiotian frontier', *American Journal of Archaeology*, 93: 274–5.

Nafissi, M. (2004). 'Class, embeddedness and the modernity of ancient Athens', *Comparative Studies in Society and History*, 46: 378–410.

—— (2005). *Ancient Athens and Modern Ideology: Value, theory and evidence in historical sources, Max Weber, Karl Polanyi and Moses Finley.* London: Institute of Classical Studies.

Neils, J. (ed.) (1992). *Goddess and Polis: the Panathenaic Festival in Ancient Athens.* Princeton: Princeton University Press.

Nelson, S. A. (1996). 'The drama of Hesiod's farm', *Classical Philology*, 91: 45–53.

Ober, J. (1985). *Fortress Attica: Defense of the Athenian land frontier 404–322 B.C.* Leiden: Brill.

Osborne, R. (1985a). 'Buildings and residence on the land in classical and hellenistic Greece: the contribution of epigraphy', *Annual of the British School at Athens*, 80: 119–28.

—— (1985b). *Demos: The discovery of classical Attika.* Cambridge: Cambridge University Press.

—— (1987). *Classical Landscape with Figures. The ancient Greek city and its countryside.* London: George Philip.

—— (1988). 'Social and economic implications of the leasing of land and property in classical and hellenistic Greece', *Chiron*, 18: 279–323.

—— (1992a). '"Is it a farm?" The definition of agricultural sites and settlements in ancient Greece', in B. Wells (ed.) *Agriculture in Ancient Greece.* Stockholm: Swedish Institute at Athens, 21–8.

—— (1992b). 'Classical Greek gardens: between farm and paradise', in J. D. Hunt (ed.) *Garden History: Issues, approaches, methods.* Washington: Dumbarton Oaks, 372–91.

—— (1996a). 'Pots, trade and the archaic Greek economy', *Antiquity*, 70: 31–44.

—— (1996b). *Greece in the Making: 1200–479 BCE.* London: Routledge.

—— (2004). 'Demography and survey', in S. Alcock and J. Cherry (eds.) *Side-by-Side Survey. Comparative regional studies in the Mediterranean world.* Oxford: Oxbow, 163–72.

Owens, E. J. (1983). 'The *koprologoi* at Athens in the fifth and fourth centuries BCE', *Classical Quarterly*, 33: 44–50.

Pansiot, F. P. and H. Rebour (1961). *Improvements in Olive Cultivation.* Rome: FAO.

Parisinou, E. (2000). *The Light of the Gods: The role of light in archaic and classical Greek cult.* London: Duckworth.

Parkins, H. and C. Smith (eds.) (1998). *Trade, Traders and the Ancient City.* London: Routledge.

Parry, J. and M. Bloch (eds.) (1989). *Money and the Morality of Exchange.* Cambridge: Cambridge University Press.

Paton, W. R. and M. J. L. (1898). 'On some Karian and Hellenic oil presses, *Journal of Hellenic Studies*, 18: 209–17.

Patterson, H. (ed.) (2004) *Bridging the Tiber: Approaches to regional archaeology in the Middle Tiber Valley.* London: British School at Rome.

Patterson, J. (1987) 'Crisis: what crisis? Rural change and urban development in imperial Appennine Italy', *Papers of the British School at Rome*, 55: 134–46.

Pellett, P. L. and Shadarevian, S. (1970). *Food Composition Tables for Use in the Middle East.* Beirut: American University at Beiruit.

Pettegrew, D. (2001). 'Chasing the classical farmstead: assessing the formation and signature of rural settlement in Greek landscape archaeology', *Journal of Mediterranean Archaeology*, 14: 189–209.

Pina-Cabral, J. (1986). *Sons of Adam, Daughters of Eve: The peasant worldview of the Alto Minho.* Oxford: Clarendon Press.

Pitt-Rivers, J. (1971). *People of the Sierra.* Chicago and London: University of Chicago Press.

Plassart, A. (1916). 'Fouilles de Délos exécutées aux frais de M. le Duc de Loubat, Quartier d'habitations privées à l'Est du Stade', *Bulletin de Correspondance Hellenique*, 40: 145–256.

Pomeroy, S. (1994). *Xenophon's Oikonomikos. A social and historical commentary.* Oxford: Oxford University Press.

Potter, T. (1979) *The Changing Landscape of South Etruria.* New York: St Martins Press.

Powell, M. (1987). 'The tree section of ur5 (HAR)-ra=hubullu', *Bulletin on Sumerian Agriculture*, 3: 145–51.

Price, S. and Nixon, L. (2005). 'Ancient Greek agricultural terraces: evidence from texts and archaeological survey', *American Journal of Archaeology*, 109: 665–94.

Prieto, A. (2005). *The 'Division Lines' of Metaponto.* Ph.D. thesis, Department of Classics, University of Texas at Austin.

Pritchett, W. K. (1956). 'The Attic Stelai, part 2', *Hesperia*, 25: 178–328.

Raftopoulos, H. (1995). *The history of Greek olive oil and perfumed oil and their containers in the Greek world between 1100 and 480 B.C.* Faculty of Literae Humaniores. D.Phil. thesis, Litt. Hum. Faculty, University of Oxford.

—— (1996). '*Η πώληση χρήση του ελαιόλαδου και του αρωματικού λαδιού στην Αρχαϊκή Ελλάδα*', in S. Papadopoulos (ed.) *Elaia kai ladhi.* Athens: ETBA (Industrial Development Bank of Greece), 110–17.

Rathbone, D. (1981) 'The development of agriculture in the Ager Cosanus during the Roman Republic: problems of evidence and interpretation', *Journal of Roman Studies*, 71: 10–23.

Rauh, N. K. (1993). *The Sacred Bonds of Commerce: Religion, economy, and trade society at Hellenistic Roman Delos 166–87 BCE.* Amsterdam: J. C. Gieben.

Reger, G. (1994). *Regionalism and Change in the Economy of Independent Delos, 314–167 BCE.* Berkeley, Los Angeles, Oxford: University of California Press.

Renfrew, C. (1972). *The Emergence of Civilisation. The Cyclades and the Aegean in the third millennium B.C.* London: Methuen.

—— and Wagstaff, J. M. (eds.) (1982). *An Island Polity. The archaeology of exploitation in Melos.* Cambridge: Cambridge University Press.

Rhodes, P. J. (1981). *A Commentary on the Aristotelian Athenaion Politeia.* Oxford: Clarendon Press.

—— (2006). 'The reforms and laws of Solon: an optimistic view', in J. Blok and A. Lardinois (eds.). *Solon of Athens: New historical and philological approaches.* Leiden: Brill, 248–60.

—— and Osborne, R. (2003) *Greek Historical Inscriptions 403–323 BCE.* Oxford: Oxford University Press.

Richards, P. (1985). *The Indigenous Agricultural Revolution: Ecology and food production in West Africa.* London: Hutchinson.

Roberts, J. T. (1994). *Athens on Trial: The antidemocratic tradition in western thought.* Princeton: Princeton University Press.

Robinson, D. M. (1929–1952). *Excavations at Olynthos*. Baltimore: Johns Hopkins University Press.

—— (1930). *Excavations at Olynthos II. Architecture and Sculpture*. Baltimore: Johns Hopkins University Press.

—— and Graham, J. W. (1938). *Excavations at Olynthos VIII. The Hellenic House*. Baltimore: Johns Hopkins University Press.

Rudolph, W. W. and Boyd, T. D. (1978). 'Excavations at Porto Cheli and vicinity, preliminary report IV: the Lower Town of Halieis, 1970–77', *Hesperi*, 47(333–55).

Runnels, C. (1990). 'Rotary querns in Greece', *Journal of Roman Archaeology*, 3: 147–54.

—— and Hansen, J. (1986). 'The olive in the prehistoric Aegean: the evidence for domestication in the early Bronze Age', *Oxford Journal of Archaeology*, 5: 299–308.

—— and van Andel, T. H. (1987). 'The evolution of settlement in the Southern Argolid, Greece: an economic explanation', *Hesperia*, 56: 303–34.

Ryder, M. L. (1983). *Sheep and Man*. London: Duckworth.

Sallares, R. (1991). *The Ecology of the Ancient Greek World*. London: Duckworth.

Saller, R. (2002). 'Framing the debate over growth in the ancient economy', in W. Scheidel and S. Von Reden (eds.) *The Ancient Economy*. Edinburgh: Edinburgh University Press, 251–69.

Salmon, J. (1983). *Wealthy Corinth: a History of the City to 338 B.C.* Oxford: Oxford University Press.

Scheidel, W. and von Reden, S. (eds.) (2002). *The Ancient Economy*. Edinburgh: Edinburgh University Press.

Scott, J. C. (1976). *The Moral Economy of the Peasant: Rebellion and subsistence in Southeast Asia*. New Haven: Yale University Press.

—— (1985). *Weapons of the Weak: Everyday forms of peasant resistance*. New Haven: Yale University Press.

Shay, J. M. and Shay, C. T. (1978). 'The plant remains', in G. R. Rapp and S. E. Aschenbrenner (eds.) *Excavations at Nichoria in Southwest Greece*. Minneapolis: University of Minnesota Press.

Shear, J. L. (2003). 'Prizes from Athens: the list of Panathenaic prizes and the sacred oil', *Zeitschrift für Papyrologie und Epigraphik*, 142: 87–108.

Skydsgaard, J.-E. (1980) 'Non-slave labour in Italy during the Late Republic', in P. Garnsey (ed.) *Non-Slave Labour in the Greco-Roman World*, Cambridge Philological Society suppl. vol. 6. Cambridge: Cambridge University Press, 65–72.

Small, A. M., Roe, B., Hayes, J. W., Simpson, C. J., Guzzetta, G., MacKinnon, M., and Monckton, S. G. (1994). 'A pit group of c.80–70 CB from Gravina di Puglia', *Papers of the British School at Rome*, 72: 197–260.

Snodgrass, A. (1987). *Archaeology in Greece: The future of a discipline.* Berkeley, New York and London: University of California Press.

Sordinas, A. (1971). *Old Olive Oil Mills and Presses on the Island of Corfu, Greece. An essay on the industrial archaeology and the ethnography of agricultural implements.* Memphis: Memphis State University Anthropological Research Center.

Spence, I. (1990). 'Pericles and the defence of Attika during the Peloponnesian war', *Journal of Hellenic Studies,* 110: 91–109.

Spurr, M. S. (1986) *Arable Cultivation in Roman Italy, c. 200 BCE–c. AD 100. Journal of Roman Studies* Monographs 3. London: Society for the Promotion of Roman Studies.

Tandy, D. W. (1997). *Warriors into Traders: The power of the market in early Greece.* Berkeley: University of California Press.

Terral, J.-F. Alonso, N., Buxó i Capdevila, R., Chatti, N., Fabre, L., Fiorentino, G., Marinval, P., Pérez Jordá, G., Pradat, B., Rovira, N., and Alibert, P. (2004). 'Historical biogeography of olive domestication (*Olea europaea L.*) as revealed by geometrical morphometry applied to biological and archaeological material', *Journal of Biogeography,* 31: 63–77.

Terral, J.-F. and Arnold-Simard, G. (1996). 'Beginnings of olive cultivation in eastern Spain in relation to holocene bioclimatic changes', *Quaternary Research,* 46: 176–85.

Thomas, K. (1983). *Man and the Natural World: Changing attitudes in England 1500–1800.* Harmondsworth: Penguin.

Thompson, D. B. (1937). 'The garden of Hephaistos', *Hesperia,* 6: 396–425.

Thompson, H. A. (1959). 'Activities in the Athenian Agora 1958', *Hesperia,* 28: 91–108.

Todd, S. (1996). 'Lysias against Nikomachos: the fate of the expert in Athenian law', in L. Foxhall and A. D. E. Lewis (eds) *Greek Law in its Political Setting: Justifications not justice.* Clarendon Press: Oxford, 101–31.

Touchais, G., Huber, S., and Philippa-Touchais, A. (2000). 'Chronique des fouilles et découvertes archéologiques en Grèce', *Bulletin de Correspondance Hellenique,* 124: 753–1023.

Tozer, H. F. (1890). *The Islands of the Aegean.* Oxford: Clarendon Press.

Tyree, L. and Stefanoudaki, E. (1996). 'The olive pit and Roman oil making', *Biblical Archaeologist,* 59: 171–8.

Valavanis, P. (1986). 'Les amphores panathénaiques et le commerce athénien de l'huile', in J.-Y. Emprereur and Y. Garlan (eds.) *Recherches sur les Amphores Grecques*. Athens: École Française d'Athènes, 453–60.

Vallet, G. (1962). 'L'introduction de l'olivier en Italie centrale'. *Hommages à A. Grenier 3*. Paris: Latomus, 1554–63.

van Andel, T. H. and Runnels, C. (1987). *Beyond the Acropolis: a Rural Greek Past*. Stanford: Stanford University Press.

von Reden, S. (1995). *Exchange in Ancient Greece*. London: Duckworth.

—— (1997). 'Money, law and exchange: coinage in the Greek polis', *Journal of Hellenic Studies*, 117: 154–76.

Walbank, M. B. (1983). 'Leases of sacred properties in Attica part IV', *Hesperia*, 52: 207–31.

Watrous, L. V. (1982). 'An Attic farm near Laurion', in *Studies in Attic Epigraphy, History and Topography. Presented to Eugene Vanderpool. Hesperia* suppl. 19. Princeton: American School of Classical Studies, 193–8.

Weber, M. (1958). *The City*. New York and London: Free Press.

Wells, B. (ed.) (1992). *Agriculture in Ancient Greece*. Proceedings of the 7th International Symposium, Swedish Institute at Athens. Stockholm: Swedish Institute at Athens.

White, K. D. (1963). 'Wheat farming in Roman times', *Antiquity*, 37: 207–12.

—— (1967). *Agricultural Implements of the Roman World*. Cambridge: Cambridge University Press.

—— (1970). *Roman Farming*. London: Thames and Hudson.

—— (1975). *Farm Equipment of the Roman World*. Cambridge: Cambridge University Press.

—— (1984). *Greek and Roman Technology*. London: Thames and Hudson.

Whitehead, D. (1977). *The Ideology of the Athenian Metic*. Cambridge: Cambridge University Press.

Wickens, J. M. (1983). 'Deinias' grave at Timesius' farm', *Hesperia*, 52: 96–9.

Wickham, C. (1988) 'Marx, Sherlock Holmes and late Roman commerce', *Journal of Roman Studies*, 78: 183–93.

Wilkins, J. M. and Hill, S. (2006). *Food in the Ancient World*. Oxford: Blackwell.

Wilkinson, T. (1982). 'The definition of ancient manured zones by means of extensive sherd sampling techniques', *Journal of Field Archaeology*, 9: 323–33.

Wood, E. M. (1983). 'Agricultural slavery in classical Athens', *American Journal of Ancient History*, 8: 1–47.

Wood, E. M. (1988). *Peasant Citizen and Slave. The Foundations of Athenian Democracy.* London and New York: Verso.

Yoshiyuki, S. (1993). 'Isolated farms in classical Athens', *Kodai*, 4: 1–19.

Young, J. H. (1956). 'Studies in South Attica, country estates at Sounion', *Hesperia*, 25: 133–44.

Young, R. S. (1951). 'An industrial district in ancient Athens', *Hesperia*, 20: 135–288.

Index Locurum

General Index

The manufacturer's authorised representative in the EU for product safety is Oxford University Press España S.A. of el Parque Empresarial San Fernando de Henares, Avenida de Castilla, 2 – 28830 Madrid (www.oup.es/en or product.safety@oup.com). OUP España S.A. also acts as importer into Spain of products made by the manufacturer.

www.ingramcontent.com/pod-product-compliance
Ingram Content Group UK Ltd.
Pitfield, Milton Keynes, MK11 3LW, UK
UKHW020846150926
475595UK00003B/26

9780198152880